Kant's concept of geography
AND ITS RELATION TO RECENT GEOGRAPHICAL THOUGHT

UNIVERSITY OF TORONTO DEPARTMENT
OF GEOGRAPHY RESEARCH PUBLICATIONS

Kant's concept of geography AND

ITS RELATION TO

RECENT GEOGRAPHICAL

THOUGHT

J. A. MAY

Published for the University of Toronto
Department of Geography
by University of Toronto Press

ISBN 0-8020-3260-5

The science of Geography . . . is, I think, quite as much as any other science a concern of the philosopher.

STRABO

There is a boundless advantage in a knowledge of the places in the world for philosophy. . . .

ROGER BACON

The revival of the science of geography . . . should create that unity of knowledge without which all learning remains only piece-work.

IMMANUEL KANT

High towers, and metaphysically-great men resembling them, round both of which there is commonly much wind, are not for me. My place is the fruitful bathos, the bottom-land, of experience.

IMMANUEL KANT

Preface

This work is a revised version of a doctoral dissertation presented to the School of Graduate Studies in the University of Toronto in December of 1967. It is the author's conviction that considerable benefit is to be derived from a detailed analysis and consideration of the thought of one of the great figures of the past, and from a comparison between that thought and contemporary thinking on some of the fundamental issues respecting the nature of geography. Kant is one of the great kaleidoscopes of Western thought.

In the analytic section of this work, more attention is paid to the concept "spatial relations" as a basic concept of the nature of geography, since it appears to me that this concept and its implications have been inadequately analyzed by geographers. On the other hand, less attention is paid to such concepts of the nature of geography as "environmentalism" and "regionalism," since these concepts have often been acutely analyzed and assessed by geographers themselves. Hence, I have only a few footnotes to add to our understanding of these concepts.

I would like to take the opportunity to express my appreciation to those who have assisted in the writing and preparation of this study, especially my advisors on the original thesis—Professor J. M. O. Wheatley of the Department of Philosophy in the University of Toronto, and Dean George Tatham of York

University—to Professor Wheatley for a painstakingly thorough job of editing the original typescript that has contributed greatly to whatever clarity of thought and ease of presentation the work may possess; to Dean Tatham whose considerable knowledge of the history of geography has saved me at times from serious blunders or omissions.

I owe debts of gratitude also to the following: to Professor Emil L. Fackenheim of the Department of Philosophy in the University of Toronto, to Mrs. Elizabeth Tunner of Toronto, and to Mrs. Eva Dawkins of Kamloops, British Columbia, for assistance in rendering Kant's sometimes intractable German into tolerable English; and to the editorial staff of the University of Toronto Department of Geography Research Publications series.

This book has been published with the help of a grant from the Social Science Research Council of Canada, using funds provided by the Canada Council. A contribution toward cost of publication has also been made by the Publications Committee of the Canadian Association of Geographers. I am most grateful for this support.

J. A. May
Toronto
March, 1970

Contents

Figures

Figures

Kant's concept of geography
AND ITS RELATION TO RECENT GEOGRAPHICAL THOUGHT

I
Introduction

KANT AND GEOGRAPHY

Immanuel Kant is the outstanding example in Western thought
of a professional philosopher concerned with geography. He
introduced the study of geography to Königsberg University in
1756 and lectured on the subject regularly for forty years, until the year before his retirement in 1797.[1] This was many
years before the first chair in geography—for Carl Ritter at
Berlin in 1820—was created at a German university. Ritter
held this chair until his death in 1859, when it was allowed to
lapse because of declining interest in geography. A second
chair in geography was not made available until 1871, at
Leipzig for Oskar Peschel. Appointments at Halle, Königsberg,
and Strassburg, and in other countries, France, Italy, Russia,
and England, followed in rapid succession.[2] Thus, Kant began

[1]See Paul Gedan's notes to Kant's "Physische Geographie," Gesammelte Schriften
(hereinafter referred to as GS), IX, 509; and Erich Adickes, Kant als Naturforscher,
II, 388.

[2]Alfred Hettner, Die Geographie: ihre Geschichte, ihr Wesen und ihre Methoden,
p. 447.

3

to lecture on geography well over one hundred years before it became an established university discipline, although he was not the first to have taught a course in geography at a German university. Adickes[3] points out that at Göttingen in 1754-55, Büsching had introduced courses on the globe and the political geography of Europe; and in the following year, J. M. Franz taught a course on the geography of North America. However, Büsching and Franz were important geographers of their day, not philosophers. For Königsberg University in 1756, geography was something entirely new.

Over the forty-year period that he lectured on geography, Kant gave his course forty-eight times. He lectured more often only on logic (54 times) and metaphysics (49 times). Next in order came moral philosophy (28 times), anthropology (24 times), and theoretical physics (20 times).[4] Thus, he devoted some seventy-two courses of lectures to the empirical sciences, geography and anthropology—a considerable portion of his teaching life. When he was appointed a professor in 1770, and his teaching load was drastically reduced from some twenty-five or thirty hours of lectures per week to approximately ten, his lectures became devoted almost exclusively to philosophical topics. Yet, he continued to lecture on geography until the end of his active teaching life.

Kant's course of lectures on geography was not a commentary on a textbook prescribed by the university or the state, as was the custom in his day. As he makes quite clear in the early announcements of his courses, he organized his own course and collected his own material from a variety of sources.[5] As late as 1788, an official decree from von Zedlitz, the Minister of Education, specifically exempted Kant's course on physical geography from the customary regulations:

---

[3] Adickes, Kant als Naturforscher, II, 388.

[4] Friedrich Paulsen, Immanuel Kant: His Life and Doctrine, pp. 57-58.

[5] In his "Entwurf und Ankündigung eines Collegii der physischen Geographie, " GS, II, 3-10, Kant points out that no suitable textbook for the course was available, and that he had started at the very beginning of his academic career to collect material from a variety of sources for the purpose of presenting geography in special lectures (p. 4). See also GS, II, 25, where Kant announces that he will again lecture on geography "from my own essays"; and GS, II, 35, where he indicates he will lecture from "my own notes. "

The worst compendium is certainly better than none, and the professors may, if they are wise enough, improve upon the author as much as they can, but lecturing on dictated passages must be absolutely stopped. From this, Professor Kant and his lectures on physical geography are to be excepted, as it is well known that there is yet no suitable text-book in this subject.[6]

Evidently, the subject held considerable personal interest for Kant. In fact, throughout his life and even in old age, his favourite reading for mental relaxation consisted of works on geography.[7]

Although his course on physical geography was given only at the introductory, first-year university level, Kant obviously attached considerable importance to it, for he regarded geography as "the propaedeutic for knowledge of the world."[8] Thus, he regarded geography as providing preliminary and essential knowledge for more advanced work. The precise sense in which Kant thought of geography as a "propaedeutic," however, remains to be determined.

Kant provides us with an interesting geographical reason for his remaining at Königsberg. Even before he had written the Critique of Pure Reason he was well known throughout Germany, and in later years could almost certainly have commanded any university appointment in philosophy in the country. In fact, pressure was put on him to accept appointments at Halle and Berlin. Kant, however, declined all offers to leave Königsberg. Undoubtedly, one of his reasons for remaining was Königsberg's ideal situation for gaining knowledge of man and of the world without having to travel. As a busy seaport, it had distinct advantages over an inland city. It was well situated for overseas trade, and for intercourse with different countries and with peoples of diverse languages and customs.[9] In addition, the Königsberg of Kant's day was a very cosmopolitan city, since it contained sizeable segments of German, Scandinavian, Dutch, English, Polish, Russian, and other Slavic populations.[10]

---

[6]Quoted in Paulsen, Immanuel Kant, p. 60.

[7]Willibald Klinke, Kant for Everyman, p. 22.

[8]"Physische Geographie," GS, IX, 157.

[9]Anthropologie in pragmatischer Hinsicht (hereinafter referred to as "Anthropologie"), GS, VII, 120-121.

[10]J. W. H. Stuckenberg, The Life of Immanuel Kant, pp. 2-4.

The foregoing biographical information indicates that geography was of considerable importance to Kant, not only academically but also personally. It was obviously more important to him than we would be led to suppose by Gerland's oft-repeated implication that Kant's concern with geography represented only an interest in empirical knowledge as necessary for his philosophical investigation of the whole of knowledge.[11] It was undoubtedly more important than many philosophers have been prepared to admit. In fact, few philosophers have paid much attention to Kant's work on geography. Kuno Fischer,[12] with his propensity to see Kant's philosophy as a doctrine of "historical development," regarded the Physical Geography as a contribution to the natural history of the earth. In support of this contention he quotes the following passage from the introduction to the Physical Geography: "It is true philosophy to trace the diverse forms of a thing through all its history." Fischer draws attention to Kant's distinction between the development of things as natural history and the "customary description of nature," although he contends that the latter "contents itself with artificially classifying things, with grouping their external attributes, and with describing what they are in their present state." Undoubtedly, this applies to one Kantian interpretation of the concept "description of nature," but Fischer ignores Kant's distinction between history and geography, and the sense in which geography as a "description of nature" differs from the "customary description of nature." Paul Menzer,[13] in his brief consideration of Kant's work on geography, was primarily concerned with an exposition of Kant's various plans for a physical geography, as presented in his Entwurf und Ankündigung eines Collegii der physischen Geographie, Nachricht von der Einrichtung seiner Vorlesungen in dem Winterhalbenjahre von 1765-1766, and in the introduction to the Physische Geographie. Menzer also draws attention to the close parallel between geography and anthropology, and to Kant's stress on the importance of geography for the practical conduct of life.

---

[11]Georg Gerland, "Immanuel Kant, seine geographischen und anthropologischen Arbeiten," Kant-Studien, X (1905), 508-509.

[12]A Critique of Kant, pp. 67-68.

[13]Kants Lehre von der Entwicklung in Natur und Geschichte, pp. 76-82.

The outstanding authority on Kant's work in geography, however, is Erich Adickes. His detailed labours gave rise to three books and two substantial chapters in a fourth.[14] In the Untersuchungen, which was specially commissioned by the Prussian Academy of Sciences for its edition of Kant's works, Adickes is primarily concerned with a detailed analysis and comparison of some twenty manuscript copies of Kant's lectures on physical geography. A brief summary of Adickes' main findings will be given in chapter iii. The Ansichten is primarily concerned with an historical survey of what Kant included in his course on geography, its historical development, the sources of his material, and a survey of his other work that could be labelled geographical. The neu Kollegheft contains Adickes' suggestions for improving the official Rink edition of Kant's Physische Geographie; it is basically a supplement to the Untersuchungen. The chapters devoted to the Physische Geographie in Kant als Naturforscher are a summary of the essentials of the other three books. Adickes, however, was primarily concerned with technical and historical questions respecting Kant's work on geography, and especially with the course of lectures on the subject that he gave for forty years. He rarely concerned himself with Kant's meaning, or with the possible relations between geography and Kant's philosophy. Hardly any attention is paid to philosophical aspects of Kant's geography. Apparently, Adickes preferred to draw a sharp line between Kant's empirical work and his theoretical work. Note, for instance, his caustic condemnation of Gerland's[15] attempt to find a connection between the introduction to the Physical Geography and the introduction to the Metaphysical Foundations of Natural Science.[16] Nevertheless, Adickes' work represents an important foundation and point of departure for any discussion of Kant's geography, and reference to some of his findings will be made in this work.

Traditionally, geographers too have paid comparatively little attention to Kant's work on geography. Some of the more traditional opinions will be examined later in the discussion of

[14]Adickes, Untersuchungen zu Kants physischer Geographie; Kants Ansichten über Geschichte und Bau der Erde; Ein neu aufgefundenes Kollegheft nach Kants Vorlesung über physische Geographie; and Kant als Naturforscher, II, 353-406.

[15]"Immanuel Kant, " pp. 500 ff.

[16]Adickes, Kant als Naturforscher, II, p. 385n.

the influence of Kant's ideas on geography. Within recent years, however, there has been a renewed interest on the part of geographers in Kant's work on geography, due primarily to the writing and influence of Richard Hartshorne and George Tatham.

Writing in 1939, Hartshorne was principally interested in Kant's work, and especially in the introduction to his Physical Geography, as providing an adequate logical foundation for geography:

Although relatively few of the students who have attempted to determine the field of geography from purely logical considerations have given this fundamental problem [the place of geography in a classification of the sciences] adequate attention, geography nevertheless stands on no weak grounds. As we have noted in our historical survey, our field received for many years the attention of one of the great masters of logical thought. In the introduction to his lectures on physical geography, Immanuel Kant presented an outline of the division of scientific knowledge in which the position of geography is made logically clear. The point of view there developed has proved so satisfactory, to others as well as to this writer, both in leading to an understanding of the nature of geography and in providing answers to all questions that have been raised, that it seems worthwhile to quote at some length from Kant's original statements. [17]

Tatham writes as follows:

. . . Kant's contribution was more philosophical . . . since it consisted of his definition of the nature of geography and its relationship to the natural sciences. This definition given in the introduction to his lectures describes so completely the scope of geography that it has affected directly or indirectly all succeeding methodological discussion. One can go further and say that confusion about the aim and content of geography has almost always only appeared when Kant's analysis has been ignored. [18]

In his article on "Geography" in the Encyclopaedia Britannica,[19] Preston James writes:

Kant, the great German master of logical thought, gave geography its place in the over-all framework of organized, objective knowledge (science). . . .

[17] Richard Hartshorne, The Nature of Geography: A Critical Survey of Current Thought in the Light of the Past (1961), p. 134.

[18] George Tatham, "Geography in the Nineteenth Century," in Geography in the Twentieth Century. ed. by Griffith Taylor, pp. 28-69; quote from p. 38.

[19] 1960 edition, X, 138-152; quote from p. 146.

8

This is essentially the concept of the place of geography among the sciences that has guided the main stream of geographic thought since Kant.

De Jong writes that, "When the chorological principle of geography is mentioned . . . we also involuntarily think of the philosopher Immanuel Kant, who in his lectures laid the foundations of geography as a chorological science."[20] No further mention of Kant is made in this book; it is just assumed that the basis of the author's point of view is to be found in Kant. More recently, Jan Broek has argued that Kant's "philosophic construction" secured for geography "an honorable status among the sciences," and moreover, that Kant's position and various restatements of it have provided "the fundamental justification for geography."[21] Hartshorne, Tatham, James, de Jong, and Broek are able representatives of the position that Kant's concept of geography provides an adequate logical foundation for the subject, or at least, failing sufficient coverage of all essential points in Kant himself, that a suitable foundation can be derived from his position.[22]

Nevertheless, these geographers all assume that Kant's logical justification for geography is based on a threefold distinction he draws between "systematic" or "subject-matter" sciences, temporal or historical sciences, and spatial or geographical sciences. It is questionable whether this distinction is Kantian. But it will have to be examined in detail when we come to consider Kant's concept of geography, and because of its importance in the recent history of geographical thought.

Also of interest are an historian's recent opinions of the importance of Kant's teaching on the interrelatedness of history and geography:

As knowledge in both domains [history and geography] grew, men of course became able to demonstrate this relationship more completely. By the last decades of the eighteenth century, the greater part of the world had been roughly mapped, while Voltaire, Hume, Gibbon, and other writers had in-

[20]G. de Jong, Chorological Differentiation as the Fundamental Principle of Geography, p. 1.

[21]Jan O. M. Broek, Geography: Its Scope and Spirit, p. 14.

[22]It is interesting also to observe that Kant has recently been granted the status of semi-official philosopher to the Education Committee of the Canadian Association of Geographers (see E. E. Owen, "The Nature of Geography," Teaching Geography, No. 10 [1964], especially pp. 1-2).

troduced the critical age in historical writing. In fine, the scientific temper
had been applied to both subjects; myths and tradition were being fast erad-
icated from cartography and historiography alike. It is not surprising, therefore, to find the philosopher Kant by this time giving a systematic statement
of the interdependence of history and geography.

A genuine understanding of this many-sided interdependence, indeed, was
perhaps first exhibited in Kant's lectures on geography at the University of
Königsberg. . . . He taught that history and geography were two parts of a
whole: one being a description of the world and its inhabitants in the order
of time, the other a description of them in the order of space. He taught
also that physical geography was the primary basis of history, for it was the
great determinant of various branches of geography which obviously affected
and controlled history. . . .

It is clear that an epochal forward step had been taken in relating the human past to the human environment when students began to be taught, for example, that the history of religions has been largely determined by soil,
climate, natural resources, physical communications, and like factors.
That teaching effectively began with Kant. [23]

There have, of course, been exceptions to the foregoing
opinions concerning Kant's importance in geography, especially
with regard to whether his position provides a suitable logical
foundation for modern geography. The notable exception is
Fred Schaefer. [24] He does not, however, deny Kant's historical significance. In fact, he indicates that the philosophical
underpinnings of Ritter, Hettner, and Hartshorne are all, more
or less, based on or derivable from Kant. The Kantian influence, however, was a disaster, since it led to the doctrine of
"exceptionalism" in geography. According to Schaefer, Kant
is clearly "the father of exceptionalism." Moreover, he "made
the exceptionalist claim not only for geography but also for history. According to him history and geography find themselves
in an exceptional position from that of the so-called systematic
sciences" (p. 232). By the doctrine of "exceptionalism,"
Schaefer is referring to the claim that certain disciplines, in
this case geography and history, are exceptions to the generally
accepted pattern of scientific explanation. For Schaefer, there
is only one form of explanation: "To explain the phenomena one

[23]Allan Nevins, The Gateway to History, pp. 302-303.
[24]"Exceptionalism in Geography: A Methodological Examination, " Annals, Association of American Geographers (hereinafter referred to as Annals, AAG), XLIII, No.
3 (1953), 226-249.

has described means always to recognize them as instances of laws" (p. 227). Thus, in order to qualify as a science, geography must abandon the "exceptionalist" position, and adopt the procedures of the natural sciences.

Although the doctrine of "exceptionalism" constitutes Schaefer's main attack on the Kantians, or at least his interpretation of Kant's position, he raises a number of other objections that may be briefly noted. Kant is depicted as "a poor geographer when compared with his contemporaries or even Bernhard Varenius who died more than one hundred and fifty years before him" (p. 232). The obvious implication here is that one has to be a good geographer in order to be qualified to discuss the nature of the subject. Schaefer goes on to argue that the distinction between the so-called systematic and the non-systematic sciences, on the grounds that the former abstract from spatial and temporal conditions, is in principle mistaken because the "systematic" sciences do not ignore spatio-temporal coordinates. Schaefer thought that Kant had written the introduction to the Physical Geography in his youth, before he had been influenced by Newton. Thus, his remarks are pre-Newtonian and pre-Critical, and it is "unfortunate so many geographers kowtow to a patently immature idea of his youth" (p. 233). He goes on to contend that geography for Kant was descriptive in the narrowest sense of that term. One can understand this historically, since Kant lived before the rise of modern social science, so that the idea of socio-historical laws was quite foreign to him. In addition, he based his notions of classification on Aristotle and Linnaeus. His "model" for geographical classification was biology, which at that time was still largely classificatory. Thus, Kant "conceived of geography exclusively as a catalogue of the spatial arrangement and distribution of taxonomic features" (p. 234). In general, Kant presents us with cosmography or natural history in its traditional sense, and not with geography at all as that discipline is understood today.

Hartshorne[25] has replied effectively to some of Schaefer's contentions. The introduction to the Physical Geography is not pre-Newtonian or pre-Critical. It dates from the mid-1770's,

[25]"'Exceptionalism in Geography' Re-Examined, " Annals, AAG, XLV, No. 3 (1955), 205-244.

not from Kant's youth. As early as 1755, his <u>General Natural History and Theory of the Heavens</u> had been written "in conformity with Newton's principles."

The contention that Kant was "a poor geographer" is most effectively met with the reply that Kant was not a geographer at all and made no pretense of being one. The body of his <u>Physical Geography</u> is not original, but was compiled from many sources, some admittedly classical even in his day, others immediately contemporary. It is thus difficult to see how Kant's geography could have been either much better or much worse than the general body of geographical knowledge of his day. The implication that one has to be a good geographer to discuss the nature of geography is sheer nonsense. Obviously, it helps considerably if the philosopher is well acquainted with the subject he is discussing, and in Kant's case there is no question that he was widely familiar with the geographical literature of his day, but to push the argument to its logical conclusion is to maintain the absurd position that most professional philosophy is meaningless. Although Schaefer's account suffers from some elementary historical errors respecting Kant, some of his points, notably those concerning abstraction, "exceptionalism," and classification, are contentious and hence worthy of serious consideration.

Schaefer's disciple, William Bunge, recently chose Kant for the following attack:

This difficulty regarding the plausibility—the intuitive reality—of theories trapped Kant. He claimed that humans are born with certain powers to discern the real from the unreal. He used as his prime example the alleged irrefutable and exclusive reality of Euclidean geometry! [26]

There appears to be no reason for discussing Kant in this context other than the fact that, following Schaefer, "positivist" geographers believe that Kant marks the spot where the geographical train went off the tracks. However, the criticism misses the mark. Kant can hardly be classified as a believer in the doctrine of innate ideas. Euclidean geometry is not valid because we possess some innate power to discern its "exclusive reality." Euclidean geometry, although a pure construction in thought, is valid for Newtonian science because the

[26]William Bunge, <u>Theoretical Geography</u>, p. 4.

12

physicist must himself make fundamentally similar construc-
tions in his own work.[27] In addition, Kant did not deny the
logical possibility of non-Euclidean geometries:

Thus there is no contradiction in the concept of a figure which is enclosed
within two straight lines, since the concepts of two straight lines and of their
coming together contain no negation of a figure. The impossibility arises not
from the concept in itself, but in connection with its construction in space,
that is, from the conditions of space and of its determination.[28]

Non-Euclidean geometries, of course, could not have qualified
as mathematics for Kant because they could not be constructed
in space. At best, they would remain mere figments of
thought.[29] Kant was not "trapped" by a doctrine of innate ideas,
but by his concept of space. However, this debate has no bearing on geography, since the space of the earth's surface that
concerns the geographer is either Euclidean or perceptual.

Lukermann,[30] in his analysis of Kant's contribution to geographic thought, appears to tread a middle path between
Hartshorne and his followers on the one hand, and Schaefer
and his followers on the other. Although Lukermann points
out that "the interests of geographers largely centred on Kant's
classifying the field of knowledge, as experienced, between
'science,' history, and geography" (p. 3), and although he evidently accepts this traditional geographical interpretation of
the threefold Kantian division of basic types of knowledge as
historically sound, he contends that

accepting the Kantian definition of space and time as a framework for geography and history, without qualification, ignores practically all the developments of nineteenth and twentieth century physics and mathematics. In
that century and a half it was shown that Kantian/Newtonian space was in
fact neither synthetic, intuitive nor self-evident, but rather logical and abstract. In essence, modern science had conceptually shifted from the intuitive three-dimensional space of Kant, directly apprehending reality, to
logical n-dimensional manifolds of mathematical space-models without direct reference to reality [p. 11].

[27]See Gottfried Martin, Kant's Metaphysics and Theory of Science, p. 36.

[28]Kant, Critique of Pure Reason, A220 = B268 (A stands for the first edition; B
stands for the second edition).

[29]Martin, Kant's Metaphysics, pp. 23-24.

[30]F. Lukermann, Geography: de Facto or de Jure, mimeographed.

This argument would seem to indicate that the traditional Kantian designation of geography as a science of space is inadequate because of post-Kantian developments of the concept of physico-mathematical space. However, Lukermann is quite unclear as to why the concept of space of contemporary astronomical physics should serve as a "model" for contemporary geography, whose concern with space is confined to the surface of the earth. Lukermann, however, does point out, although critically, that such contemporary "macroscopic geographers" as Schaefer, Warntz, Stewart, and Isard accept the view that "Space is independent of the phenomena it contains—seemingly a fundamental concept to science since Newton, to philosophy since Kant and, therefore, predicate in a modern scientific geography" (p. 5).

The foregoing illustrations of recent geographical interest in Kant reveal differences in meaning. interpretation, and significance. Evidently, Kant's work on geography requires more attention than either philosophers or geographers have yet been prepared to give it.

One may ask five general questions concerning Kant's philosophy of geography:

(1) What are the origins of Kant's ideas on geography, or are they original?
(2) What development did Kant's ideas on geography undergo in his own thinking?
(3) What influence did Kant's ideas on geography have?
(4) What does Kant mean?
(5) To what extent can Kant be regarded as an adequate foundation for modern geography?

Questions (1) to (3) will be discussed in an introductory manner in chapter iii. Although the broad lines of development of Kant's thought on geography are fairly clear, satisfactory answers to questions concerning origins and influence would probably require years of detailed historical research to unravel. By origins, of course, I am referring to the possible sources of Kant's ideas on geography, especially as they are expressed in the introduction to the Physical Geography, and not to the main body of his lectures. This material is not original, and we are well acquainted with Kant's sources, due mainly to the painstaking research of Erich Adickes and Paul

Gedan.[31] Kant himself, in his earliest announcement of his
course on physical geography, was quite explicit concerning
his major sources.[32] These sources included Varenius' Geographia Generalis, Buffon's Histoire Naturelle, Lulofs' Einleitung zu der mathematischen und physikalischen Kenntniss der
Erdkugel, current scientific journals from Göttingen, Hamburg,
and Leipzig, the papers of the scientific academies of Paris and
Stockholm, and travelogues of the day. Other major sources,
added to the course from time to time, included Linnaeus' Systema Naturae, Leibniz' Protogaea, Woodward's An Essay towards a Natural History of the Earth, Whiston's A New Theory
of the Earth, Halle's Naturgeschichte der Thiere, Buache's
Essai de géographie physique, Bergman's Physikalische Besch-
reibung der Erdkugel, and Büsching's Neue Erdbeschreibung.
On the other hand, the question of the possible sources of
Kant's ideas on the nature of geography, particularly as expressed in the introduction to the Physische Geographie, has
never been adequately explored.

   This study will concentrate on questions (4) and (5). It is
essential to get at Kant's meaning. Philosophers, aside from
Adickes, have paid little attention to Kant's geography, and as
indicated earlier, Adickes shows scant interest in the question
of meaning. Geographers, especially in the last century, paid
slight attention to Kant's introduction to the Physische Geographie. And those who studied the body of that work inevitably
found it sadly lacking, for there was nothing unusual or original in it. Today, the content of Kant's course on physical
geography, although occasionally informative, is little more
than an historical curiosity. Consequently, it is my intention
to concentrate on Kant's ideas on geography, on his philosophy
of geography. However, geographers who have paid attention
to Kant's introduction to Physische Geographie have not sought
elsewhere in his works for interpretations of the meaning of
some of the terms he uses. The Physische Geographie was an
introductory university course, in which Kant did not go deeply
into the meaning of his terms. It is necessary, therefore, to

---

[31]A résumé of these sources may be found in Gedan's notes to the "Physische Geographie," GS, IX, 552-568.
[32]"Entwurf und Ankündigung eines Collegii der physischen Geographie," GS, II, 4.

look at other contexts to determine more precisely Kant's meaning and use of such important terms as: "world," "whole," "man," "nature," "description of nature," "experience," "knowledge," "system," "idea," "architectonic," "propaedeutic," "outer and inner sense," "history," "space," "time"— all of which occur in the introduction to Physische Geographie. The question concerning Kant's meaning will be dealt with in chapters iv, v, and vi. The text of the introduction to Physische Geographie, however, is based on students' lecture notes and therefore it should probably not be pushed too far respecting some specific meanings and uses. One must be mindful of Kant's own shrewd observation on students' lecture notes:

> Those of my listeners who are able to grasp the gist correctly, take down the lectures least explicitly, least like a dictation, but jot down merely the main points, to ponder over them afterwards. Those who are explicit in their transcript, seldom have the power of judgement to distinguish what is important from what is unimportant, and pile up heaps of misunderstood stuff among what they might have grasped correctly. [33]

Nevertheless, Adickes' research into some twenty manuscript copies of the Physische Geographie indicates that they are in essential agreement on the main points respecting Kant's ideas on the nature of geography.

The approach to be taken in this study involves the assumption that a complete separation cannot be made between Kant's empirical work and his theoretical work, especially with respect to his ideas or thoughts on the empirical disciplines, geography and anthropology. If geography is to serve, in some sense, as a propaedeutic, then it seems reasonable to assume that common Kantian terms used in the philosophy of geography bear a close relation to the same concepts as employed in more theoretical contexts. At the same time, however, it must also be borne in mind that the distinction between empirical and theoretical is an important one for Kant. This proviso suggests that a special twist may be given to some of these common philosophical terms as they are employed in empirical contexts.

Question (5), concerning the extent to which Kant can be regarded as an adequate foundation for modern geography, will

---

[33] Letter to Marcus Herz, October 20, 1778, Kant, trans. and ed. by Gabriele Rabel, p. 103.

be approached from the point of view of an analysis of contemporary positions on some major issues respecting the nature of geography, and a comparison of these with Kant's position. It is not possible to deal with the whole content of a field as broad as geography, nor even with all its conceivable philosophical issues. Rather, some major issues will be selected that have been recurrent themes in the history of geographical thought, that concerned Kant, and that are also of central importance in geography today.

These themes include: (1) the limits and scope of geography; (2) concepts of what geography is about, or more generally, geography as a science; and (3) the place of geography in a classification of the sciences. Within the context of these general themes, more specific issues of importance will require discussion. Something needs to be said on the concept "region," since one of the basic points of departure for geography, as a science of the earth's surface, is the fact that that surface is not uniform but manifoldly differentiated into "regions," whatever meaning one wishes to attach to the concept "region." Another issue that requires discussion is "explanation in geography," since maintenance of the traditional threefold distinction between "systematic," historical, and geographical sciences depends logically, in part, on the possibility of demonstrating that there is a distinctive geographical form of explanation. Geographers, on the whole, have paid little explicit attention to the question of explanation, although Lukermann[34] has made it a central issue in several short papers. "Positivist" geographers, such as Schaefer and Bunge, have treated the issue of explanation largely in a polemical manner, through their insistence that geography, in order to qualify as a science, must follow the explanatory example of the advanced physical sciences. Finally, something will be said on the relations between geography and history, not only because of the traditional association between the two disciplines, but also because of the importance of historical explanation in some areas in geography. These themes, although they will be isolated to some extent for purposes of analysis, are all to some degree interrelated and

---

[34]Lukermann, "On Explanation, Model, and Description," The Professional Geographer, XII, No. 1 (1960), 1-2; and "The Role of Theory in Geographical Inquiry," ibid., XIII, No. 2 (1961), 1-6.

interdependent and consequently, there will be no attempt to draw rigid lines between them. Thus, some repetition and overlapping between chapters is unavoidable.

The questions to be discussed do not have definitive answers. It would be foolish to suppose that basic issues, some of which have been recurrent themes for centuries, can in any strict sense be resolved, but discussion of them may achieve some further clarification. In addition, this work is not conceived basically as a contribution to the history of geographical thought, except with respect to what light it may shed on Kant's concept of geography. In the chapters on contemporary geography, the orientation will be primarily analytic.

PHILOSOPHY AND GEOGRAPHY

Throughout the history of Western thought, comparatively little attention has been paid to geography by philosophers. Kant is one of the outstanding exceptions to this generalization. There appear to be two prima facie reasons for this situation.

In the first place, man has generally been more concerned with time than with space, as it affects his immediate existence. Man, of all creatures inhabiting the earth, is the only one that knows that he must die, and thus he is the only one that can conceive his own end. And so, time in this existential sense becomes a baffling and sometimes terrifying notion. As a natural extension of his immediate concern with time, man the philosopher has been concerned with history, with the record of man's "progress" in time. On the other hand, modern man has generally been less concerned with his immediate relations to space; he has tended to take them for granted. As a result, "philosophy of geography" is an odd-sounding phrase, whereas "philosophy of history" is not.

Not all peoples, however, possess this notion of the primacy of time over space. Among most primitive nomadic peoples, spatial relations are more basic than temporal ones. For instance, among the Nuer, within the repetitive annual cycle of events, itself determined by natural changes, particular daily occurrences such as milking-time and meal-times tend to be arranged not in an objective temporal sense, but as

a result of certain coordinated activities in the spatial movements of a group of people. Although daily events follow a logical sequence, they are not controlled by an abstract and
objective temporal system, since there are no autonomous
points of reference to which activities should conform. Thus,
it is milking-time because the Nuer have gathered to milk their
cattle, not because it is "time" to milk cattle.[35] Similarly,
the Eskimo, because of their tendency to treat all events as
contemporaneous, apparently lack the ability to give a careful
chronological account of events. This tendency is reflected in
the Eskimo language, which is concerned, verbally, with position and spatial relations and not with tenses. In short, "the
importance we lavish on time, Eskimo accord to space."[36]
But once man became sedentary, and his position in space was
relatively fixed, the need to adopt some system that measured
time objectively became apparent. For instance, the earliest
surviving document of known authorship on Chinese agriculture,[37] dating from the first century B.C., although it contains
some reference to calculation of time in terms of the occurrence
of certain natural events on the earth's surface,[38] is striking
because of the number of references to time calculated in an
objective sense, as based on knowledge of the solstices.[39]

In present-day North American society, we tend to temporalize many of our basic spatial notions. The following pair
of brief conversational exchanges may illustrate the point:

"How far away is x?" "Oh, about fifteen minutes."
"How long will it take to get to x?" "Oh, about a mile."

The first exchange, provided that the means of getting to x is
understood, is immediately intelligible in our society; the second exchange is not. This tendency to temporalize spatial no-

---

[35]E. E. Evans-Pritchard, "Time and Space," The Nuer, chap. iii, pp. 94-138.

[36]Edmund Carpenter, Frederick Varley, and Robert Flaherty, "Eskimo," Explorations, IX (1959).

[37]Fan Shêng-Chih, On "Fan Shêng-Chih Shu"; An Agriculturistic Book of China Written in the First Century B.C., trans. by Shih Shêng-Han.

[38]For instance, "Light soils are to be ploughed when apricot-trees come in blossom.
Plough again when the blossom fades. . . . Never plough too early. Wait till grasses
sprout" (ibid., p. 7).

[39]Ibid., pp. 11, 15, 19, 27, 37.

tions is also quite apparent in professional literature. For instance, R. D. McKenzie, one of the early leaders of the "human ecology" movement, interprets the concept "distance" temporally:

> Ecological distance is a measure of fluidity. It is a time-cost concept rather than a unit of space. It is measured by minutes and cents rather than by yards and miles. By time-cost measurement the distance from A to B may be farther than from B to A, provided B is upgrade from A.[40]

Nevertheless, there are indications that man is becoming more concerned with his relations in the space of the earth's surface. Some of these indications are the greatly increased mobility of contemporary man, concern over rapidly growing urban sprawl, the need to preserve valuable agricultural land and to plan more carefully the use and arrangement of geographical space, and the world "population explosion" and the attendant need to find adequate space for a variety of meaningful human activities on the surface of our already crowded planet. We may yet come to agree with Henry Miller that "more than anything they [human beings] need to be surrounded with sufficient space—space even more than time."[41] These multiplying concerns, and the growing vitality of contemporary geography itself, may lead eventually to greater philosophical concern with geography and its problems.

The second underlying reason for the lack of philosophical interest in geography is attributable to the rather obscure academic position occupied by geography. As a science, it tends to maintain a somewhat peripheral position, seemingly dependent upon a variety of materials that are the concern of more specialized disciplines. In many university calendars, one may find geography listed under both the science and the arts sections. It thus appears, in some vague sense, to cut across certain readily recognized distinctions. Its very diffuseness seems to defy precise definition. Outsiders generally know little about geography, but appear to agree that it is concerned with place names and location, and, more elegantly, with that ill-defined and all-encompassing notion, "the environment."

[40]R. D. McKenzie, "The Scope of Human Ecology," in The Urban Community, ed. by Ernest W. Burgess, pp. 167-182; quote from p. 170.

[41]Henry Miller, Tropic of Cancer (New York: Grove Press, 1961), p. 318.

In fact, many social scientists came to identify "the environment" as "the geographical factor." The problem has been
rather neatly expressed by one geographer as follows:

Most people would be hard put to it to say exactly what geography is. Botany is the study of plants; ichthyology the study of fish; mineralogy the study
of minerals; physics is a little more difficult, it shades off into mathematics.
But on the whole people know, or at least feel that they could find out, what
the natural scientists study. More would hesitate to define history and many
would dispute any definition given, but history is an old and eminently respectable subject and everyone knows what it is about. But geography . . . ?
Some look upon the geographer as a kind of intellectual rag-and-bone man
content to cull ill-assorted bits and pieces of information from many other
disciplines. The bookshelves and the book lists of the geographer might perhaps seem to support the idea that he is a dilettante dabbling in many things,
mastering none.[42]

This tendency to be eclectic has meant that geography has run
the risk, from time to time, of disappearing as a serious academic discipline. At times it has been regarded as little
more than the "hand-maiden of history." At other times, during periods of crisis, it has tended to be absorbed by the natural sciences, the social sciences, and history. During such
periods of crisis, when its philosophical underpinnings have
become insecure, it has even disappeared from the academic
scene. In addition, geography's position as a "following"
rather than as a "leading" discipline has meant that philosophers, on the whole, have tended to pay little attention to it;
they have preferred to pay attention to the more basic ideas
that geography has acquired from other disciplines.

Geography is nevertheless a very old study. It is truly,
along with history and philosophy, one of the "mother" sciences
from which other, more specialized, sciences have broken off
and developed. Although the concept "geography" was not invented by Eratosthenes until the third century B.C.,[43] the first
clearly recognizable geographical work is the Periegesis of
Hecataeus of Miletus, written in the sixth century B.C.[44]

[42]J. B. Mitchell, Historical Geography, p. 1.

[43]C. van Paassen, The Classical Tradition of Geography, pp. 44–45.

[44]For an account of the geography of Hecataeus, see E. H. Bunbury, A History of
Ancient Geography, I, pp. 134–148; originally published, 1883. A more recent and
more detailed account is contained in Lionel Pearson, Early Ionian Historians (1939),
pp. 25–106. Pearson discusses the Periegesis in detail (pp. 34–96). He indicates

Hecataeus had the reputation of being "a much-travelled man, " and he recorded his descriptions of the places and peoples in Europe and Asia that he had visited. However, his work, along with that of others, was condemned by Heraclitus as a "mere collection of disparate and unrelated facts. "[45] Thus, the debate as to whether geography should be descriptive or explanatory is a very old one indeed.

These are some of the reasons why professional philosophers have, on the whole, been little concerned with geography. On the other hand, geographers have often been concerned to demonstrate that their discipline is philosophically supportable and respectable. Geography, in part because it has largely been ignored by professional philosophers, has to some extent developed its own philosophical tradition, although this tradition has never really been divorced from major philosophical positions. The classical example of this geographical concern with philosophy is Strabo. He introduces his work[46] by remarking that geography should be "quite as much as any other science, a concern of the philosopher. " He gives three reasons: (1) all the great geographers were also philosophers; (2) geography, like philosophy, requires wide learning, possessed solely by the man who has investigated "things both human and divine"; and (3) because of its aim and utility, geography is akin to philosophy; both investigate "the art of life, that is, of happiness. " Strabo closes his introduction with the remark: "I have said thus much to show that the present work is a serious one, and one worthy of a philosopher" (1, 1, 23).

that this work was properly geographical and not an historical inquiry, and that "it is as a pioneer in geographical inquiry that he [Hecataeus] was most famous" (p. 28). Earlier, J. B. Bury, although pointing out that Hecataeus was "first and foremost a geographer, " had regarded him also as "initiat[ing] the composition of 'modern' history" (The Ancient Greek Historians, pp. 11-12; originally published, 1908). George Sarton, A History of Science, I, pp. 184-188, discusses Hecataeus as "the father of geography. " An exception to this generalization, however, is provided by W. A. Heidel, "Anaximander's Book, The Earliest Known Geographical Treatise, " Proceedings of the American Academy of Arts and Sciences, LVI (1921), 239-288, who obviously regards Anaximander as the first geographer.

[45]G. S. Kirk and J. E. Raven, The Presocratic Philosophers, p. 189. Much later, Strabo also expressed his displeasure with Hecataeus, "unscientific" handling of both geography and history (The Geography of Strabo, Bk. 7, chap. 3, sec. 6; Bk. 8, chap. 3, sec. 9).

[46]The Geography of Strabo, 1, 1, 1.

Several philosophical positions have more recently played important roles in providing the orientation and underpinnings for geography. Exploring the place of Kantianism in geographical thought will be one of the major undertakings of this work. German idealism was certainly influential in the thinking of Carl Ritter, who is widely regarded as one of the founders of modern geography. Nineteenth-century materialism and determinism were very influential in German geography during the latter half of that century, especially in the thought of prominent figures such as Oskar Peschel and Georg Gerland. This influence extended well into the twentieth century, and can be seen in the thinking of important geographers such as Ellsworth Huntington and Griffith Taylor. However, probably the greatest philosophical influence on twentieth-century geographical thought so far has been exerted by the neo-Kantianism of Wilhelm Windelband and Heinrich Rickert. Their influence on some aspects of the thought of Alfred Hettner [47] was of considerable importance, and Hettner's great influence on Richard Hartshorne, probably the dominant American philosopher of geography of this century, is too well known to require comment.

Especially in Western Europe, varieties of "phenomenology" appear to be having some influence on basic orientation in geography. For instance, note van Paassen's interesting general observation:

Geographical science has in fact a phenomenological basis; that is to say, it derives from a geographical consciousness. On the one hand, the geographer develops this consciousness and makes society more aware of geography, but on the other hand the rise of geographical science is dependent upon the existence of a pre-scientific and natural geographical consciousness. Just as we can only practice the science of history if we experience history, and just as there can only be historians and a historical science in a society with a historical sense, so can geographers and geographical science exist only in a society with a geographical sense. The history-less robot society and the 'brave new world' in which the deliberate falsification of history preserves the last shreds of the historical sense, is at the same time a geography-less society, for both time and space have been reduced to mere dimensions. [48]

[47] For Hettner's acknowledgment of his debt to Windelband and Rickert, see his Die Geographie, pp. 112-114. This statement, however, requires the qualification and enlargement to be made later.

[48] Van Paassen, The Classical Tradition, p. 21.

This tendency may also be seen in contemporary French geography, in the interest some geographers have shown in the ways that people perceive and experience local space and spatial relations.[49] It may also be seen, although perhaps less clearly, in the preoccupation of some contemporary German geographers with local landscapes. Insofar as these units of local landscape are looked upon as only physico-topographical or physico-biotic-topographical, they are regarded as the smallest objective homogeneous topographical units that can be discovered. However, the concept "Soziotop" appears to involve human feelings for, perception of, and identification with a local landscape.[50] In the United States, perhaps the best representative of this orientation in geography is David Lowenthal,[51] especially as regards his concern with the "uniqueness of private milieus," "personal variations in aspects of world views," and "subjective elements in private geographies."

Finally, particularly in Sweden and the United States, "positivism" has been making considerable headway recently. Positivism in geography may be briefly defined as the position that maintains we can classify as science only that which can employ advanced mathematical techniques, translate procedures into such techniques, and predict on the basis of mathematically formulated laws; and we must ignore all else.[52]

Professional philosophical interest in geography does not begin with Kant, nor of course does it end with him. A brief historical review of philosophy of geography will reveal some of the major recurrent issues.

[49]See Georges Matoré, L'Espace Humaine (Paris, 1963).

[50]For a recent review of this German literature, see de Jong, Chorological Differentiation, pp. 64-74.

[51]"Geography, Experience, and Imagination: Towards a Geographical Epistemology," Annals, AAG, LI, No. 3 (1961), 241-260.

[52]A recent, if somewhat extreme, representative of this position is Bunge, Theoretical Geography.

II
Philosophy and geography, an historical sketch

This brief historical review assumes a philosophical orientation. Geography shall be viewed primarily from the perspective of what mention has been made by professional philosophers
of that discipline. Furthermore, there will be little concern
with specific geographical content, although it may often be
found in the writings of philosophers, but rather with material
that gives rise to discussion of the nature of geography.

THE CLASSICAL PERIOD

Reference to geographical matters begins very early in the history of Western philosophy, with the Presocratics.[1] Thales apparently discussed the causes of the annual flooding of the Nile
River. Anaximander is generally credited with having been the
first known cartographer of record, and with having initiated

---

[1]Detailed discussion of the points mentioned in this paragraph may be found in Kirk
and Raven, The Presocratic Philosophers.

the study of meteorological and climatological phenomena. Xenophanes could be called the first "Neptunist," since he believed that the ocean was the begetter of all water, clouds, winds, and rivers, and indeed of the earth itself. The Pythagoreans were probably the first to have regarded the earth as a sphere, essential for the establishment of geodesy, and to have divided the earth's surface into the traditional five zones, one torrid or equatorial, two temperate, and two frigid or polar. The Presocratic philosophers, however, regarded nature as a single unity. Although they often touched upon geographical matters, the slicing up of nature into segments, and the recognition of one of these segments as the concern of geography, would never have occurred to them. Thus, they present us with a philosophy of nature, but in no sense with a philosophy of geography. Only in retrospect can one isolate certain segments of their work and label it geography.

With the coming of Socrates, philosophical interest centres on man, although not to the total neglect of nature. With this reorientation in philosophy, we might expect to find some reference to questions of human geography. In fact, the beginnings of economic geography can perhaps be seen in Xenophon's pamphlet, Ways and Means.[2] In this work, Xenophon was concerned with setting forth proposals for improving the economic life of Athens. He begins with a geographical survey of the physical features, climate, and natural resources of Attica. He remarks on the ideal situation of Athens relative to the inhabited world, its neighbours, and land and sea routes. He points out the physical and economic advantages of Athens as a seaport, and dwells at length upon the economic conditions of the silver mines at Laurium and the possibilities of expanding the benefits to be derived from further exploitation of this natural resource. Plato indicates that the legislator should take into account a variety of geographical factors, and frame his laws accordingly, for differences in soil, climate, and water have their effects upon men.[3] He believed that the Greeks enjoyed a geographical situation which was exceptionally favour-

---

[2]Contained in The Greek Historians, ed. by Francis R. B. Godolphin, II, pp. 644-657.

[3]Plato, Laws, Bk. 5, 747.

26

able for the attainment of excellence, since Greece lay "midway between winter and summer."[4] He seems also to have believed that the inhabited portion or ecumene of the earth, as known to the Greeks, was far more extensive than "those who are wont to describe" the earth (i.e., geographers) were prepared to admit.[5] Moreover, Plato's writings contain perhaps the most striking account of denudation and soil erosion, and their implications for civilization, to be found in ancient literature.[6] But again, in neither Xenophon nor Plato is there any indication that a separate science which could be labelled geography exists.

Aristotle's work, aside from Heraclitus' remark on Hecataeus of Miletus, appears to have been the first philosophy that gave rise to debate over the nature of geography, although the "debate" is really nonexistent from his side since the concept "geography" had not yet been invented. Aristotle was probably the first to offer a classification of the sciences. He divided the sciences into three broad categories: theoretical, practical, and productive. Although all science can be said to aim at knowledge in some sense, the theoretical sciences aim at knowledge for its own sake, the practical sciences at knowledge as a guide to conduct, and the productive sciences at the creation of some useful or beautiful object. The theoretical sciences, in turn, are divided into three classes: physics, mathematics, and first philosophy, only the first of which concerns us. "Physics deals with things that exist separately but are not immovable,"[7] i.e., with natural objects capable of motion and change. Physics proper deals with natural bodies in general, with the principles and causes of motion and rest that apply to all natural physical objects. However, in the introduction to his Meteorologica,[8] Aristotle indicates that although nature should be regarded as a unity, it can be subdivided for purposes of study. In moving from the most general to the more

[4]Epinomis, 987 A, D (if Plato wrote the Epinomis).

[5]Phaedo, 108-109.

[6]Critias, 111 A-D.

[7]Aristotle, Metaphysics, 1026a.

[8]338a-339a. Quotations in this section are taken from The Works of Aristotle, ed. by W. D. Ross. Meteorologica is in Vol. III.

particular, Aristotle begins by considering physics proper, "the first causes of nature, and all natural motion." Next, he considers the motions of the stars, or astronomy, and "becoming and perishing in general." Aristotle then indicates that a part of the inquiry into nature concerns what had come to be called meteorology, which "is concerned with events that are natural, though their order is less perfect than that of the first elements of bodies. . . . Of these things some puzzle us, while others admit of explanation in some degree." Meteorology covers a broad array of natural occurrences, beginning with phenomena such as the milky way, comets, and meteors, which Aristotle assumed to take place "just below the sphere of the moon," then descending to atmospheric phenomena and their effects such as rain, cloud, mist, dew, hoarfrost, snow, hail, and wind. Meteorology is also concerned with occurrences on the surface of the earth, such as rivers, springs, and floods, and finally with events that occur within the earth, i.e., earthquakes. Aristotle points out that the study of physics (of nature) will not be complete until an account of animals and plants (of biology) is given. Biology, of course, is a category of natural phenomena quite distinct from meteorology, yet it is to be studied "in accordance with the method we have followed" so as to preserve the unity of the study of nature.

Aristotle did not recognize geography as a distinct branch of the study of nature. Although the four elements, fire, air, water, and earth, interact with one another materially, in considering meteorological phenomena, "fire occupies the highest place and earth the lowest." In short, the major efficient cause of natural events on the earth's surface is the sun. But, we only possess unqualified scientific knowledge of something when we know the cause on which it depends.[9] Thus, it appears that no separate natural science that confined itself to the earth's surface would be possible for Aristotle, since so many of the things on the earth's surface depend upon causes that originate outside or beyond that surface.

Strabo enters this debate with the following remark:

So much for Poseidonius. For in my detailed discussions many of his views will meet with fitting criticism, so far as they relate to geography;

---

[9]Aristotle, Posterior Analytics, 71b8-12.

28

but so far as they relate to physics, I must inspect them elsewhere or else not consider them at all. For in Poseidonius there is much inquiry into causes and much imitating of Aristotle—precisely what our school [the Stoic] avoids, on account of the obscurity of the causes.[10]

For Strabo, geography is evidently not physics. Geography is concerned more with effects than with causes. He puts his position even more clearly when he remarks: "But he [the geographer] should take some other things on faith, even if he does not see a reason for them; for the question of causes belongs to the student of philosophy alone."[11] And when it comes to discussing the distribution of various phenomena on the surface of the earth, Strabo indicates quite clearly that the geographer is concerned with phenomena that are "accidental" and due to "chance":

. . . a distribution of animals, plants, and climates as exists is not the result of design—just as the differences of race, or of language, are not, either—but rather of accident and chance. And again, as regards the various arts and faculties and institutions of mankind, most of them, whence once men have made a beginning, flourish in any latitude whatsoever and in certain instances even in spite of the latitude; so that some local characteristics of a people come by nature, others by training and habit.[12]

For Aristotle, however, there could be "no science of the accidental":

. . . that there is no science of the accidental is obvious; for all science is either of that which is always or of that which is for the most part. . . . But that which is contrary to the usual law science will be unable to state. . . the accidental is contrary to such laws. We have stated, then, what the accidental is, and from what cause it arises, and that there is no science which deals with it.[13]

Thus, Aristotle would not have recognized geography, as conceived by Strabo, as a separate and distinct science. Strabo's debate with Posidonius and Aristotle introduces a couple of very old and basic philosophical issues in the history of geographical thought—"What kind of a science is geography?"; and "What is

[10]The Geography of Strabo, 2, 3, 8.

[11]Ibid., 1, 1, 21. See the rest of 1, 1, 21 and 1, 1, 20, for an enumeration of these "other things" Strabo believes the geographer should "take on faith."

[12]Ibid., 2, 3, 7.

[13]Aristotle, Metaphysics, 1027a.

the place of geography in a classification of the sciences?" We shall return to these themes in greater detail in later contexts.

Since the basic philosophical issue has been covered, we need not linger over Aristotle's specific contributions to geography, although they were considerably more extensive than those of any of his predecessors. Aristotle, however, also touched upon matters of human geography, and thus his work may be thought of as containing an implicit distinction between physical and human geography. Following Hippocrates and Plato, he indicates that the Greeks are "intermediate in geographical position," and therefore unite within themselves the qualities of peoples living in colder and hotter climates.[14] Aristotle also wrote extensively on geographical aspects of planning. For instance, he discusses the quantity and distribution of population in the ideal state, the moderate size of its territory, the proper economic relations of the central city to its surrounding countryside, the merits of having close connections with the sea, and the planning of the central city itself in terms of exposure to sun and climate, access to a good water supply, topographic considerations of defence, convenience of communications, and the preservation of beauty.[15] One can, in retrospect, find these geographical matters in Aristotle. He would not, of course, have recognized them as separable, but treated them as intimate aspects of "politics," the practical art of governing the state.

The classical period of Greek philosophy was characterized by a conception of the unity of nature, although, following Socrates, human activity tended to be separated out from nature and treated as a distinct category. The philosopher was the "wise man," the universal intellect, who inquired into all things. Specialization was foreign to the classical Greek mind. The coming of the Hellenistic Age, however, marked a period of specialization, during which the majority of men ceased to inquire into everything or into the whole, and thus ceased to be philosophers in this traditional sense. In fact, Eratosthenes

[14]Aristotle, Politics, 1327b.

[15]Ibid., 1325b-1327b, 1330a-1331a. Plato had earlier touched upon aspects of the size of the ideal state (Republic, Bk. 5, 459; Laws, Bk. 5, 737-738). Cf. Whittlesey on geographical aspects of the ideal state (Derwent Whittlesey, The Earth and the State, pp. 22-23).

was possibly the first to have called himself a scientist as distinct from a philosopher.[16] During this period, two philosophers made important contributions to geographical thought— Eratosthenes and Posidonius.

Although Eratosthenes[17] would not be regarded today as a philosopher of note, apparently he did write a history of philosophy and several commentaries on Plato. Strabo refers to his ethical and rhetorical treatises, On the Good and Studies in Declamation. In the same context, however, Strabo offers the opinion that Eratosthenes was "constantly vacillating between his desire to be a philosopher and his reluctance to devote himself entirely to this profession, and who therefore succeeds in advancing only far enough to have the appearance of being a philosopher; or of the man who has provided himself with this as a diversion from his regular work, either for his pastime or even amusement."[18] However insecure Eratosthenes' status as a philosopher may be, there is no question concerning the fact that he was the first great mathematical geographer or geodesist. By "geography," Eratosthenes probably meant literally "the drawing of the earth." He conceived geography in a basically mathematical and cartographic sense, as a natural science closely allied to astronomy, and thus formulated one of the basic definitions of the field that has persisted through the ages. The link with astronomy was based on the fact that accurate measurement of distances on the earth's surface and accurate determination of location are dependent upon the earth as an astronomical body and upon its position and motion relative to other such bodies. Eratosthenes also wrote descriptions of various countries, although one recent commentator, van Paassen,[19] has indicated that these descriptions cannot be regarded as regional geography; they were "maps in prose," bare cartographic frameworks that in no way attempted a regional synthesis of data.

---

[16]Eduard Zeller, Outlines of the History of Greek Philosophy, p. 208.

[17]For recent discussion of Eratosthenes as a geographer, see van Paassen, The Classical Tradition, pp. 33-53; and Sarton, A History of Science, II, 99-114.

[18]The Geography of Strabo, 1, 2, 2.

[19]The Classical Tradition, pp. 50-51.

Posidonius is a somewhat enigmatic figure in philosophy.[20]
If one accepts the point of view that he was a systematic thinker
who held a consistent, monistic account of the world, then the
most that can be said is that Posidonius was the first to have
based a discipline recognizable as geography on a consciously
held philosophical position. Evidently following Aristotle in
part, he divided the domain of "meteorology" into three realms:
astronomy proper, the atmosphere, and earth and water and
their relations. The realm of earth and water and their relations, or the surface of the earth, is the most concrete and
empirical. Because of Posidonius' concern with concrete forms
and empirical diversity, his "meteorology" had to terminate in
a geography of the earth's surface. Yet, to preserve the unity
of his system of thought, he had to demonstrate that the empirical diversity occurring on the earth's surface could be explained in terms of astronomical and atmospheric forces. He
achieves this in part in his work, On the Oceans, where he
demonstrates, probably for the first time, that the tides are
dependent upon the joint action of the sun and the moon. Yet,
this work also contains his discussion of the division of the surface of the earth into seven zones instead of the usual five. He
reasoned that the equatorial region has a milder temperature
than the torrid, desert zones to the north and south of it. This
deduction is again based on astronomical and atmospheric evidence, but Posidonius apparently added these two zones out of
an interest in human activity, from the fact that the way of life
was strikingly different in these desert regions. A "tension"

[20]For accounts of Posidonius as a geographer, see Bunbury, Ancient Geography, II,
93-100; and Sarton, A History of Science, II, 417-418. The most detailed recent discussion is to be found in van Paassen, The Classical Tradition, pp. 332-358. However, van Paassen bases his interpretation of Posidonius as a philosopher primarily
on Reinhardt's account. Reinhardt apparently regarded Posidonius as the last great
universal mind and systematic thinker of Hellenism, one who viewed the world as an
organism and the earth as the visible form of an all-embracing cosmic force (pp.
346-348). A second tradition of interpretation has maintained, however, that the ex-
tant fragments of Posidonius give the distinct impression of an eclectic and enigmatic
figure, content to assimilate unrelated ideas from his predecessors, notably Plato,
Aristotle, and the early Stoics. It may be that van Paassen reads more systematic
unity into Posidonius than the evidence warrants, although he is careful to point out
that "any attempt which is made to characterize the geography and ethnography of
Posidonius must be imperfect, because there is insufficient direct knowledge of his
work" (p. 356). However, it is beyond the scope of this study to enter this controversy.

is introduced into his "system" by his consideration of man as the bridge between nature and the divine. This view of man undoubtedly accounts for Posidonius' great interest in human activity, and in the migration and distribution of peoples, but it appears to introduce a dualism into his so-called monism. For instance, he apparently maintained that the character of a people and advances in civilization are not due to man's response to natural conditions he encounters, but to philosophy, to ideas in the minds of men. [21] The spurious Aristotelian work, De Mundo, [22] which may have been based on two works by Posidonius, is interesting since it illustrates an attempt to overcome this basic dualism. The author draws attention to the great diversity of phenomena on the earth's surface—verdure, mountains, rivers, animals, cities, and the other works of man (392b)—yet he pities those who are content to describe merely "a single region or the plan of a single city or the dimensions of a river or the scenery of a mountain" (391a). These diverse phenomena are to be explained in terms of "divine nature," yet the author finally concedes that "wherefore the earth and the things upon the earth, being farthest removed from the benefit which proceeds from God, seem feeble and incoherent and full of much confusion" (397b). In a sense, this offers an "explanation" for the diversity of phenomena, but at the expense of failure to maintain a strong, monistic position. In fact, Posidonius often resorted to idiographic descriptions of human activity and regional differences. In the "system" of Posidonius, there appears to be an underlying tension between diversity and unity, between the concrete-idiographic on the one hand, and the organic-teleological-mystical on the other hand. He is of considerable interest in the history of geographical thought, however, since his work appears to illustrate, for the first time, a basic and recurrent problem in geography—that of resolving the dualism between factual description of diverse phenomena occurring on the surface of the earth, and adequate theoretical frameworks for explaining the diversity.

[21]See Seneca, Letter 90, "Philosophy and Progress," The Stoic Philosophy of Seneca, trans. and ed. by Moses Hadas, pp. 226-238.

[22]Contained in The Works of Aristotle, III. This work should not be pushed too far since its authorship is uncertain.

If, on the other hand, one adopts the point of view that Posidonius was not a systematic but an eclectic thinker, the least that can be said about his geography is that he was "the most intelligent traveller in antiquity,"[23] probably the greatest physical, as distinct from mathematical, geographer of the ancient world; and that he did significant and suggestive work in human geography, especially as regards ethnography, the migration and distribution of peoples. His influence on Strabo, despite the latter's occasional philosophical disagreements with him, was very great. Much of what is best in the physical geography sections of Strabo probably came from Posidonius, since Strabo often refers to Posidonius in these sections.

FRANCIS BACON AND CONTEMPORARIES

Undoubtedly the most important figure in a philosophical history of geography, from the time of Posidonius to that of Kant, is Francis Bacon, despite the fact that he rarely refers to geography as such. A strong case can be made for the claim that Bacon was most impressed with the geographical discoveries of his age, and that these discoveries served as the basis for his inspiration and indicated to him that the "wisdom of the ancients" could be surpassed. As he remarks on this theme:

And surely, when I set before me the condition of these times, in which learning hath made her third visitation or circuit in all the qualities thereof . . . the noble helps and lights which we have by the travails of ancient writers; the art of printing, which communicateth books to men of all fortunes; the openness of the world by navigation, which hath disclosed multitudes of experiments, and a mass of natural history . . . I cannot but be raised to this persuasion that this third period of time will far surpass that of the Grecian and Roman learning.[24]

In another context, Bacon makes his point even more succinctly:

But to circle the earth, as the heavenly bodies do, was not done or enterprised till these latter times. . . . And this proficience in navigation and

---

[23]H. F. Tozer, History of Ancient Geography, rev. by M. Cary, p. 190.

[24]Francis Bacon, The Advancement of Learning, p. 208. For a discussion of this point, see Benjamin Farrington, Francis Bacon: Philosopher of Industrial Science, pp. 40-43.

discoveries may plant also an expectation of the further proficience and augmentation of all sciences; because it may seem they are ordained by God to be coevals, that is, to meet in one age. [25]

Bacon classified "the parts of human learning" with reference to the three parts of man's understanding—"history to his memory, poesy to his imagination, and philosophy to his reason" (p. 69). He generally employs the term "history" in a very broad sense, more equivalent to the original Greek meaning of "historia" or "factual inquiry into" than to our contemporary, restricted meaning of a reference to events occurring in past time. When used in this broad sense, the term "history" includes much that can be recognized as geography. Bacon does, of course, employ the term in the more restricted sense in connection with some of the many types of history that he recognized. For instance, the type of "just and perfect history" which represents "a time" is called "a chronicle" (p. 74). Bacon indicates that history of nature is of three sorts—"nature in course which is history of creatures," "nature erring or varying which is history of marvels," and "nature altered or wrought which is history of art" (p. 70). The idea of a "history of marvels" comes closest to geography, and is perhaps a reference to the "Wonders and Curiosities" which formed a prominent part of works in geography of Bacon's day. [26] Bacon, however, goes on to condemn these "fabulous experiments and secrets, and frivolous impostures for pleasure and strangeness," and indicates that much investigation of nature is needed to reject these "fables and popular errors." He expresses his position with respect to these "curiosities of nature," and their place in serious scientific inquiry, more clearly in his <u>Preparative toward Natural and Experimental History</u>, [27] when he remarks

. . . that superfluity of natural histories in descriptions and pictures of species, and the curious variety of the same, is not much to the purpose. For small varieties of this kind are only a kind of sports and wanton freaks

---

[25] <u>The Advancement of Learning</u>, pp. 79-80. Unless otherwise indicated, page references in this section are to this volume.

[26] See E. G. R. Taylor, <u>Late Tudor and Early Stuart Geography</u>, p. 85.

[27] Contained in F. Bacon, <u>The New Organon and Related Writings</u>, ed. by Fulton H. Anderson, pp. 269-292; quote from p. 275.

of nature and come near to the nature of individuals. They afford a pleas-
ant recreation in wandering among them and looking at them as objects in
themselves, but the information they yield to the sciences is slight and almost superfluous.

Of most interest in his discussion of the "history of marvels"
is the contrast he draws between "singularities of place and
region" and "the strange events of time and chance" (p. 70).
This reference appears to be an implicit recognition of a basic
distinction between geography and history, areal differentiation
of phenomena on the earth's surface as contrasted with changes
through time. However, the type of history which may be recognized most clearly as geography is "history of cosmography, "
which Bacon defines as "history manifoldly mixed, " since it
is composed of irregularly conjoined elements which cannot be
strictly separated by man. History of cosmography is "com-
pounded of natural history, in respect of the regions themselves;
of history civil, in respect of the habitations, regiments, and
manners of the people; and the mathematics, in respect of the
climates and configurations towards the heavens" (p. 79). Evidently, then, cosmography relates to natural events occurring
on the surface of the earth, human habitation and customs, and
climate and atmospheric phenomena that must be mathematically determined. It is in the context of this discussion of the
"history of cosmography" that the passage referring to the
great geographical discoveries, quoted above, occurs.

Bacon throws further light on geography in his Preparative
toward Natural and Experimental History. He divides the "history of generations" into five parts (pp. 276-277)—ether and
things celestial; meteors and the regions of air which lie between the moon and the surface of the earth; earth and sea; the
four elements: fire, air, water, and earth; and finally species,
or natural history in its principal employment, wherein metals,
minerals, fossils, plants, fishes, birds, insects, and animals
are investigated. Bacon indicates, however, that these categories are to some extent artificial and a matter of convenience,
for the general consideration of the four elements "enters into
the second and third [categories], as they are integral parts of
the world. " In the appendix to the Preparative toward Natural
and Experimental History, "Catalogue of Particular Histories
by Titles" (pp. 285-292), we can recognize several of his histories as pertaining primarily to the third category, earth and

36

sea. These include a history of seasons and temperatures, which is to consider variations according to both the regions of the earth and time over periods of years, in addition to floods, heats, and droughts, etc.; a history of earth and sea in terms of their shape, size, and configurations, in addition to a consideration of islands, gulfs, salt lakes, isthmuses, and promontories; a history of greater motions and perturbations, including earthquakes, the birth of new islands, encroachments and inundations and recessions of the sea, and geysers; a history of ebbs and flows, currents, and motions of the sea; and a history of "accidents" of the sea including saltiness, colours, depths, and mountains and valleys under the sea. However, of greatest interest here is the fact that Bacon recognizes one category of history as geography proper, or geography in the strict sense, which he restricts to the land (ge = earth). The "Natural History of Geography" includes "Mountains, Valleys, Woods, Plains, Sands, Marshes, Lakes, Rivers, Torrents, Springs, and every variety of their course, and the like; leaving apart Nations, Provinces, Cities, and such like matters pertaining to Civil life" (p. 286). Of the 130 natural histories Bacon lists in his "Catalogue, " geography is the only one he feels called upon to label "natural, " which indicates that he is leaving out one aspect of the subject. Geography, as we have already seen, is intimately concerned with "Nations, Provinces, Cities, and such like matters pertaining to Civil life. " Thus, geography displays the unusual quality of cutting across the distinction between natural and civil. This quality of geography has been a recurrent and troublesome theme in the history of geographical thought.

In The Advancement of Learning Bacon goes on to point out that "natural science or theory is divided into physique and metaphysique" (p. 91). Physics and metaphysics are concerned with the inquiry into causes, whereas natural history is not.

Physique . . . is situate in a middle term or distance between Natural History and Metaphysique. For natural history describeth the variety of things; physique, the causes, but variable or respective causes; and metaphysique, the fixed and constant causes [p. 93].

Natural history then is fact-finding, concerned with the variety and particularity of things, yet it is intimately connected with physics as an essential aspect of that science.

Physique hath three parts; whereof two respect nature united or collected, the third contemplateth nature diffused or distributed. . . . The third is the doctrine concerning all variety and particularity of things . . . whereof there needeth no enumeration, this part being but a gloss, or paraphrase, that attendeth upon the text of natural history [pp. 93-94].

Knowledge forms an hierarchical pyramid for Bacon, and the essential broad basis for the hierarchy is natural history; metaphysics, as generalized physics, forms the apex.

. . . for knowledges are of pyramids, whereof history is the basis. So of natural philosophy, the basis is natural history; the stage next the basis is physique; the stage next the vertical point is metaphysique [p. 95].

In another context, Bacon enlarges upon the crucial importance of natural history:

. . . let such a history be once provided and well set forth, and let there be added to it such auxiliary and light-giving experiments as in the very course of interpretation will present themselves or will have to be found out, and the investigation of nature and of all sciences will be the work of a few years. This, therefore, must be done or the business must be given up. For in this way, and in this way only, can the foundations of a true and active philosophy be established; and then will men wake as from deep sleep, and at once perceive what a difference there is between the dogmas and figments of the wit and a true and active philosophy, and what it is in questions of nature to consult nature herself. [28]

Geography, as a significant branch of natural history, thus plays an important role in the "consultation" of nature. As natural history it is fact-finding and descriptive, but nevertheless one of the basic sciences, since the advance of physics and metaphysics depends in part upon the foundations laid down by geography.

Perhaps more typical of the continental view of geography's place among the sciences at this time is Hobbes' consideration of the matter. [29] Hobbes divided knowledge into two general categories: "knowledge of fact, " and "knowledge of the consequences of one affirmation to another. " Only the latter can be called "science"; the former comprises "history. " Geography as a science is a part of cosmography, which studies the "con-

---

[28]F. Bacon, Preparative toward Natural and Experimental History in The New Organon and Related Writings, p. 272.

[29]Thomas Hobbes, "Of the Several Subjects of Knowledge, " Leviathan, Bk. I, chap. ix, 53-55.

sequences from the motion and quantity of the greater parts of the world, as the earth and stars. " The celestial part of cos- mography is astronomy, and the terrestrial part is geography. Hobbes gives no indication that geography is also to be consid- ered as part of "history. "

This uncompromising division of geography into two dis- tinct kinds of knowledge can be seen quite clearly in the follow- ing century in D'Alembert's classification of the sciences. In that classification, geography is defined as the "mathematical description of the earth, " where earth is equivalent to land, since hydrography is the "mathematical description of waters. " D'Alembert, however, reserves the term "geology" or "sci- ence of the continents, " which is a branch of physics, for the inquiry into the causes of terrestrial physical phenomena.[30] At the same time, D'Alembert also regards geography as one of the two major "offshoots and supports" of "History, " the other being chronology. Chronology "locates men in time, " whereas geography "distributes them over the globe. " The former is the science of "time, " the latter the science of "place. " Nevertheless, D'Alembert indicates that despite the fact "both draw considerable help from . . . historical facts and celestial observations, " they are fact-finding sciences (p. 35).

ENVIRONMENTALISM

Environmentalism, that is, the study of man in relation to his physical environment, has been a recurrent theme throughout the entire history of Western thought. In its earliest formula- tions, the doctrine tends to stress the effects of physical envi- ronmental factors on various human activities and institutions. Its first striking employment occurs in the fourth or fifth cen- tury B.C., in the Hippocratic work, On Airs, Waters, and Places. We have already mentioned its influence on Plato and Aristotle. In the late middle ages, rather extreme forms of the

[30]Jean D'Alembert, Preliminary Discourse to the Encyclopedia of Diderot, pp. 144- 145, 153-154.

doctrine are to be found in Ibn Khaldûn's The Muqaddimah,[31] and Roger Bacon's Opus Majus.[32] In the sixteenth century in Europe, the doctrine was revived in a somewhat less extreme form in Jean Bodin's Six Books of the Commonwealth.[33] It probably reached its peak of popularity in the eighteenth century in Montesquieu's The Spirit of the Laws,[34] and can also be found in Rousseau.[35] Perhaps the most extreme statement of the issue, as a thoroughgoing environmental determinism, is the position adopted by the French philosopher, Victor Cousin:

Yes, gentlemen, give me the map of a country, its configuration, its climate, its waters, its winds and all its physical geography; give me its natural productions, its flora, its zoology, and I pledge myself to tell you, a priori, what

[31]See especially the third, fourth, and fifth "prefatory discussions, " I, chap. i, 167-183, where Ibn Khaldûn discusses such matters as "the influence of the air upon the color of human beings and upon many other aspects of their condition, " "the influence of the air upon human character, " and "differences with regard to abundance and scarcity of food in the various inhabited regions and how they affect the human body and character. "

[32]See especially I, 320, where Bacon comments generally on the importance of a knowledge of the "places" and "regions" of the earth; ". . . in accordance with the diversity of places is the diversity of things; and not only is this true in the things of nature, but in those of morals and of the sciences, as we see in the case of men that they have different manners according to the diversity of regions and busy themselves in different arts and sciences. . . . This attestation consists in a knowledge of the longitude and latitude of every place; for we shall then know under what stars each

place is . . . and from what planets and signs places receive their control, all of which things cause the different characteristics of places. " Leaving aside astrological implications, it is evident that for Bacon the diversity of human customs and institutions is to be explained in terms of astronomical and atmospheric forces, and of the climates and physical features that result on the earth's surface.

[33]See especially Bk. V, chap. i, 145-157. However, Bodin points out that "this compulsion [the natural inclination of peoples] is not of the order of necessity." Although physical environmental factors condition "natural inclinations" and human behaviour, their influence can be modified. The wise legislator must have a thorough knowledge of these "inclinations" for he "must know when and how to overcome, and when and how to humour these inclinations" (p. 157).

[34]See especially I, Bk. XIV, "Of Laws in Relation to the Nature of the Climate, " 221-234. Again, as in the case of Bodin, Montesquieu regards the physical environment as conditioning rather than determining human character, since he argues, for instance, that the good legislator will oppose the "vices of the climate, " and remarks in this context that "the more the physical causes incline mankind to inaction, the more the moral causes should estrange them from it" (p. 226).

[35]For instance, see The Social Contract, Bk. II, chaps. ix-x, contained in The Social Contract and Discourses, 37-41. Note especially Rousseau's implication that the boundaries of states should correspond with the boundaries of climates, and his remark on the greater fertility of mountain women.

the man of this country will be, and what part this country will play in history, not by accident but of necessity, not at one epoch but at all epochs.[36]

Opposition to excessive statements on "environmentalism," which emanated mainly from the continent, came principally from Britain. The earliest philosophical statement may have been that of Francis Bacon, who apparently was writing against the growing influence of "environmentalism" as expressed in Bodin's Commonwealth, which had been translated into English in 1606. In discussing variations among the lives of men in different parts of the world and in different nations, Bacon concludes by remarking that "this difference comes not from soil, not from climate, not from race, but from the arts."[37] The most concerted opposition to "environmentalism" generally, however, was expressed by David Hume. In his essay, "Of National Characters,"[38] he remarks that: "As to physical causes, I am inclined to doubt altogether of their operation in this particular; nor do I think that men owe any thing of their temper or genius to the air, food, or climate" (pp. 205-206). This is because "physical causes have no discernible operation on the human mind" (p. 209). Hume presents a number of concrete illustrations to refute the more extreme views of the "environmentalists," and poses the general question, "Is it conceivable that the qualities of the air should change exactly with the limits of an empire, which depends so much on the accidents of battles, negotiations, and marriages?" (p. 210). Differences in national character are due to "moral causes," including the nature of government, changes in public affairs, the plenty or poverty in which a people live, and the situation of a nation with respect to its neighbours. "A nation is nothing but a collection of individuals," and the effects of "moral causes" will vary from individual to individual (p. 203). In opposition to continental thinkers, who believed that laws and forms of government must be adapted to national character, which is in turn shaped by environmental factors, Hume believed that national character was itself shaped by these legal and political agencies. Hume, of course, was not always com-

[36]Quoted in S. W. Wooldridge and W. Gordon East, The Spirit and Purpose of Geography, p. 32.

[37]F. Bacon, The New Organon, Bk. I, aphorism 129.

[38]Contained in Hume's Essays Moral, Political and Literary, pp. 202-220.

41

pletely oblivious to the broad possibilities of geographical influence. For instance, he remarks in his essay "Of the Rise and Progress of the Arts and Sciences":[39]

> If we consider the face of the globe, Europe, of all the four parts of the world, is the most broken by seas, rivers, and mountains, and Greece of all countries of Europe. Hence these regions were naturally divided into several distinct governments; and hence the sciences arose in Greece, and Europe has been hitherto the most constant habitation of them.

But there is little in philosophical debate before Kant which indicates that the study of man in relation to his physical environment is the concern of geography. Most philosophers appear to have followed Aristotle's lead in regarding such study as an aspect of "politics." On the other hand, Strabo certainly implied that this type of study was properly the concern of the geographer, and Ptolemy[40] may have included it explicitly in geography. Strabo and Ptolemy, however, were geographers, not philosophers. An exception to the above generalization is perhaps provided in the cases of Ibn Khaldûn and Roger Bacon.[41]

Following Kant, especially in Germany, there was a tendency to regard "environmentalism" as a part of geography. For instance, Herder's Ideas towards a Philosophy of the History of Man contains several introductory volumes of almost purely geographical material, and he often discusses man in relation to his physical environment. Nevertheless, one commentator[42] has recently pointed out that Herder often employs

[39]Ibid., pp. 112-138; quote from pp. 123-124.

[40]Tetrabiblos, Bk. II, chap. 2, 120-127. The Tetrabiblos, however, is a work in astrology, not geography. Nevertheless, Sarton (Ancient Science and Modern Civilization, p. 63) summarizes Bk. II of the Tetrabiblos as being concerned with "astrological geography and ethnography."

[41]The first five "prefatory discussions" of Ibn Khaldûn's The Muqaddimah are clearly geographical, and were probably included to provide knowledge of the physical environmental background of man necessary for a comprehensive understanding of human history. See Franz Rosenthal's introduction to his translation of The Muqaddimah, I, lxxiin, where he comments on the similarity between the ideas of Ptolemy and Ibn Khaldûn. The matter is less clearly seen in Roger Bacon, who, also following Ptolemy, included geography under the section of his Opus Majus devoted to mathematics, despite the fact that much of it consisted of little more than regional descriptions of various places and peoples. Nevertheless, Roger Bacon's geography also contains discussion of man in relation to his physical environment. For discussion of Roger Bacon's geography, see John H. Bridges, The Life and Work of Roger Bacon: An Introduction to the "Opus Majus," pp. 153-156.

[42]F. M. Barnard, Herder's Social and Political Thought, pp. 121-122.

the term "climate" so broadly that he identifies it with the total milieu, human as well as physical, into which an individual is born. Thus, he presents us with an organic view of the relations between physical and human factors, and not with a doctrine of environmental causation.

Hegel's most important thoughts on the topic are to be found in his "Geographical Basis of History."[43] Since the earth is the stage upon which the "World Spirit" plays its part, the geographical basis, conceived as natural environment, is essential and necessary. Different natural environments give rise to different characteristics in peoples. Hegel, however, was no simple environmental determinist. As he points out: "Nature should not be rated too high nor too low: the mild Ionic sky certainly contributed to the charm of the Homeric poems, yet this alone can produce no Homers" (p. 80). Rather, the geographical basis, although presenting definite limits, also presents different peoples with a variety of "special possibilities," depending upon the type of natural environment. Nature may nevertheless dominate man. Thus, Hegel rules out the frigid and torrid zones as being of no historical importance. In these environments, the forces of nature set severe limits on what man can achieve; in fact, man is faced with a constant struggle with nature, and never succeeds in rising above it. As a result, peoples living in the frigid and torrid zones have never succeeded in creating civilizations of world-historical importance. Perhaps Hegel's most striking illustration of the "geographical basis of history" is his reference to the role played in ancient history by the Mediterranean Sea. It joined the nations surrounding its shores into intimate union, facilitated easy communications and the spread of ideas in a way that land masses with their mountain barriers could never have done, and indeed it made "World History" as we know it possible.

The Mediterranean is thus the heart of the Old World, for it is that which conditioned and vitalized it. Without it the History of the World could not be conceived: it would be like ancient Rome and Athens without the forum, where all the life of the city came together [p. 87].

---

[43]G. W. F. Hegel, The Philosophy of History, pp. 79-102.

One nevertheless gets the distinct impression that both Herder and Hegel thought of geography primarily as supplying background information necessary for a thorough understanding of human history. Thus, they appear to perpetuate a tradition that regarded geography essentially as "the hand-maiden of history."

Philosophical debate on "environmentalism" continues in the twentieth century. Morris Cohen's[44] examination of the problem is largely historical, and he tends to side with the French geographical school of Paul Vidal de la Blache and Jean Brunhes, rejecting the extremes of environmental determinism and "idealism." All environmental deterministic accounts are facile oversimplifications that focus attention on one set of phenomena to the exclusion of human factors. On the other hand, historical accounts which concern themselves with only human motivations and actions are equally overfacile. Cohen believes that "history must take a macroscopic view of human events,"[45] and that geographical factors are necessary conditions of historical explanation. Thus, "it is only in the interaction of man and his environment that the basic elements of history can be found" (p. 171). History and geography are inextricably bound up with one another. Cohen concludes his account with the following general observation:

To say that anything is determined by its environment is actually to say only that in order to explain anything we must look to its relations with other things—which amounts to an undeniable tautology. The essential fact is that the environment of every human being and the context of every human act contains human and nonhuman elements inextricably intertwined. Only as we realize that the events of human history include both mind and matter as polar components can we escape the grosser errors of those who would spin the world out of ideas and those who look to earth, air, fire, and water to explain all human phenomena [p. 171].

Recently, the British philosopher A. C. Montefiore, in collaboration with the geographer W. M. Williams,[46] has subjected to critical analysis the ideas of environmental determinism and possibilism in geography. The authors conclude

---

[44] The Meaning of Human History, chap. v, "The Geographic Factor in History," pp. 133-171; Reason and Nature, pp. 336, 343; American Thought, pp. 53-55.

[45] The Meaning of Human History, p. 169.

[46] "Determinism and Possibilism," Geographical Studies, II (1955), 1-11.

generally that since most geographers today agree that crude "environmentalism" is dead, and since the issue has virtually no bearing on the activities of working geographers, then debate has reached a point of no further return. Determinism versus free will is a metaphysical dispute and its solution is a question of philosophical faith. Thus, it has no direct bearing on the practical activity of geographers.

THE NEO-KANTIANS

Despite the general influence of neo-Kantianism on Alfred Hettner, the only philosopher of this school who discussed geography to any extent was Heinrich Rickert.[47] He was primarily concerned with the question of where geography fitted into his broad classification of sciences into natural and cultural. Geography may fall within either class, depending upon the point of view taken towards the subject matter in a particular study. If natural features of the earth's surface are related to cultural life, or treated as the scene of cultural evolution, then these features acquire an interest beyond their status as mere natural objects, and geography becomes a cultural science. If, on the other hand, features of the earth's surface are viewed simply as products of nature, then geography is a natural science (p. 23). Rickert expands his discussion in a later context (pp. 131-132). He recognizes three types of geographical concepts, or three modes of concept-formation in geography. Insofar as the earth's surface is regarded as the theatre of cultural evolution, then geographical conditions are necessary for the existence and development of culture, and the formation of individual geographical concepts is governed by general cultural values. Particular features of the earth's surface acquire individuality because of the relation between their unique peculiarities and the history of culture. Rickert proceeds to draw a distinction between geography and geology. In the formation of general theories, rivers, oceans, mountains, and other physical features of the earth's surface are regarded only as instances that exemplify generic types. This type of general study is called geological and not geograph-

[47]Science and History.

ical. Thus, Rickert appears to regard physical geography as an exclusively idiographic discipline, whereas geology is its nomothetic counterpart. But this theory presents difficulties. Rickert goes on to point out that a third kind of concept-formation, "that by which it [geography] achieves individual representations of definite parts of the earth's surface that do not stand in any relation to culture," cannot be accommodated within his system of classifying sciences. It is not altogether clear what Rickert means by the phrase quoted above, although it appears to refer to the study of physical features within specific regional contexts. Rickert "solves" the problem by denying that this aspect of geography can qualify as a science. Such studies can properly be regarded only as mere "compilations of data," comparable to maps and other geographical representations which also constitute mere scientific data, awaiting further conceptual elaboration through their relevance to either historical-cultural studies or generalizing theories. Thus, at least one aspect of geography falls rather awkwardly between Rickert's division of science into natural and cultural. It remained for Hettner to attempt to solve this problem, partially along neo-Kantian lines.

THE TWENTIETH CENTURY

Speaking very generally, twentieth-century philosophy has been characterized by a decline in interest in traditional all-encompassing metaphysical schemes, and by a growing interest in a variety of specific, analytic techniques and approaches. As we have seen from the foregoing pages, many traditional philosophers, with their expansive interest in the whole of reality, touched upon geography to a greater or less degree, since it studied an important aspect of reality. Little of this interest, however, can be labelled analytic, in the sense that few traditional philosophers reveal much detailed or critical concern with questions pertaining to the kind of science geography is, or to its place in a classification of the sciences, etc. On the other hand, contemporary analytic philosophers, insofar as they have been concerned with science, have shown next to no interest in geography. The advanced physical sciences have been largely accepted as the prototype of all respectable science. Recently, however, professional philosophers have been displaying a growing interest in problems peculiar to the biological sciences, the social sciences, and notably history. But philosophers of history, with one exception (Morris Cohen), have shown little inclination to explore the relations between history and geography, or the role of geography in historical explanation. There are, however, some exceptions to the generalization that twentieth-century philosophers have shown next to no interest in geography.

Probably the most notable exception, since it deals with a wide range of geographical issues, is an article by the Austrian philosopher, Viktor Kraft.[48] Although it is short, its main points are worth a brief review. Kraft begins by drawing attention to some of the conflicting definitions and conceptions of geography held by geographers themselves. He points out that Alexander von Humboldt and Carl Ritter had achieved a regeneration of the "science of the earth." Ritter paid particular attention to the place of man in geography. Peschel, on the other hand, grounded geography on a general, natural scientific point of view. This tendency was carried to its logical extreme by Gerland, who excluded man from geography. Richthofen attempted to restore a balance in geography by defining it as the science of the earth's surface to be achieved through a study of the causal interrelation of phenomena on that surface. And Ratzel returned man to a position of central importance in geography (p. 3). Kraft also draws attention to the wide range of scientific disciplines generally to be found under the name "geography." He identifies ten distinct areas of study: regional geography and the political geography of nations, wherein the object is to describe and explain individual regions or political units; geomorphology; oceanography and hydrography; climatology; plant and animal geography; anthropogeography, a broad category that includes human geography generally, made up of settlement geography, economic geography, geography of transportation, and the distribution of man and his various institutions, cultural, economic, political, religious, etc., over the face of the earth; historical geography,

[48]"Die Geographie als Wissenschaft," in Enzyklopädie der Erdkunde, Vol. I: Methodenlehre der Geographie, pp. 1-22.

whether natural, cultural or political; cartography; astronomical or mathematical geography and geodesy; and aspects of geophysics (pp. 3-4).

Leaving man aside for the moment, Kraft divides these branches of geography into three broad categories: mathematical-astronomical, physical, and biological. However, astronomical geography, geodesy, and geophysics deal with the earth-body as a whole; thus, although they form the theoretical basis for aspects of geography, logically they are not parts of geography but special "chapters" of astronomy or astrophysics, those pertaining to the earth. Geography, in order that it may be distinguished from other earth sciences, must logically be regarded as the science of the earth's surface, not of the earth as a whole. Kraft goes on to point out that stones, plants, animals, and men, which form objects of study for particular sciences, are also objects of study for geography (p. 4). Although he does not explicitly make the point, the implication is that geography cannot be distinguished from other sciences on the basis of the objects it studies, since these objects are all claimed by more specialized disciplines.

Kraft next raises the issue of the place of man in geography (pp. 5-6). In one sense, he rejects the idea that there is a fundamental dualism between physical and human geography. Geography, insofar as it considers man, does not consider him from purely political, religious, etc., points of view. In this sense geography is not split between mutually exclusive physical and human perspectives. Rather, geography's concern is with the mutual relations between man and the earth. Thus, geography, as human geography, regards the earth as the "home of man," although Kraft immediately points out that this notion does not encompass the whole of geography; it is not the point of view of geomorphology and climatology.

Kraft proceeds to discuss briefly the status of these two disciplines within geography (pp. 6-7). His conclusion with respect to the logical position of both within a general body of knowledge clearly recognizable as geography tends to be negative. He indicates that geomorphology appears to represent the tail-end of historical geology, and falls between geology and geography. Analogously, climatology is represented as a common meeting-ground (gemeinsames Arbeitsgebeit) for geology and meteorology. Although the inclusion in geography of ge-

omorphology and climatology cannot be justified on strictly
logical grounds, since they fail to fall within uniform and
clearly fixed scientific spheres, nevertheless geographers are
prominently active in both disciplines. Thus, there is some
justification for the pragmatic argument that "geography is what
geographers do."

Kraft next discusses the distinction between general or systematic geography and individual or regional geography (pp. 9-
13). Although a dualism between physical and human geography
is rejected, that between "systematic" and regional geography
appears to be a fundamental dualism inherent in the nature of
the subject. Nevertheless, this conclusion is mitigated by the
observation that geography is essentially chorological, and
therefore regional geography forms the core. If climatology
depends on meteorology for its basic principles, geomorphology
on geology, and biogeography on biology, etc., then these "systematic" aspects of geography may logically be regarded as not
purely geographical, but as auxiliary sciences (Hilfswissen-
schaft) whose aim is to help describe and explain aspects of regional geography. In addition, Kraft observes that because of
the rapid development of various sciences today, geomorphology,
hydrography, etc. may very well develop as fully independent
sciences. In this case, true geography would be constituted by
regional geography alone.

These are the essential points in Kraft's paper. He goes
on to discuss geographical methods (pp. 13-20), and concludes
with some observations on the view that regional geography is
essentially a form of art, and not a science (pp. 20-22). He
rejects this view. Art and science are diametrically opposed
perspectives. Regional geography must employ acceptable scientific techniques and procedures. There are few issues in
Kraft's paper that had not been raised and discussed by geographers themselves, especially Hettner. Nevertheless, he
brings up, even if briefly, some crucial and controversial issues relating to the nature of geography that will have to be
discussed in greater detail in later contexts.

Henryk Mehlberg,[49] within his broad classification of sciences into law-finding and fact-finding, considers geography to
be a descriptive and fact-finding science (p. 92), concerned

[49]The Reach of Science.

essentially with "local events"; its subject matter "consists of single, particular facts, rather than of general propositions expressing pervasive regularities of the universe" (p. 60). He also discusses briefly the question of scientific significance and value in geography (pp. 20-21). A geographer, in describing the surface of the earth, would consider too detailed a description to be scientifically insignificant. A mountain range deserves the geographer's attention, an anthill does not. Thus, inevitably, the geographer must abstract and generalize. Basically, he must rely on his sense of values to decide what is geographically significant.

Although he has apparently never written anything on geography, Gustav Bergmann deserves mention because of his influence on the "positivist" geographers, Fred Schaefer[50] and William Bunge.[51] In fact, he could almost be regarded as a co-author of Schaefer's article since the latter collaborated with him, and since he edited the article and saw it through the press following Schaefer's death.

In concluding this historical survey of philosophical interest in geography, brief mention should be made here of Alfred Hettner. Although a geographer, he received considerable academic training in philosophy, and has done work of high calibre in the philosophy of geography. Hettner chose philosophy as the subsidiary subject for his doctoral examination, and his interest in that discipline was so great that for a time he was torn between geography and philosophy as an academic career.[52] Since he is a significant contemporary figure in geography, some of his ideas will be discussed later in this work.

[50]"Exceptionalism in Geography," 226.

[51]Theoretical Geography, p. viii.

[52]Hettner, "From My Life," referred to in H. Schmitthenner, "Alfred Hettner," Geographische Zeitschrift, XLVII (1941), 441-468, trans. by Department of Geography and Anthropology, Louisiana State University, mimeo., p. 9.

III

The origins, development, and influence of Kant's concept of geography

ORIGINS, SIMILARITIES, AND DISSIMILARITIES

A solution to the problem of the origins or sources of Kant's ideas on geography would require years of detailed historical research. In this chapter Kant's position on only some of the major issues of geographical thought will be indicated briefly and his position compared with those of some of his predecessors. (Kant's position on these matters will be examined in greater detail in the next three chapters.) We will be concerned here primarily with the views of geographers rather than with those of philosophers.

Many of Kant's ideas on geography can be found, in embryonic or undeveloped form in the works of J. M. Franz, a geographer active in the period from 1740 to 1770.[1] Typically, Kant might have found in Franz's work all that was needed to spark his imagination to develop the ideas more fully. Yet he never refers to Franz, nor does Rink in the numerous bibliog-

---

[1] For this insight, I am indebted to Dean George Tatham, York University.

raphical references he added to Kant's Physical Geography.[2] Of course, since the ideas are not well developed by Franz, the influence could have been quite subtle, and what may appear today as a fairly obvious link need not have presented itself to Kant as consciously acknowledgeable. Hartshorne remarks that: "The earliest definite statement of the comparison of geography and history that I have found is that of J. M. Franz in 1747."[3] However, Hartshorne goes on to point out that Franz, in one of the ways typical of his period, conceived geography as closely related to astronomy, as well as to history. Both geography and astronomy were included under the common term, cosmography. In fact, the first geographical society in Germany, which Franz founded, was called "Die Cosmographische Gesellschaft." It thus appears that Franz thought of geography, at least in part, in terms of one of its basic definitions, as a mathematical science closely allied to astronomy.

Two traditions of basic definition or conception of the field of geography have persisted through the ages. The first, as we have seen, stems from Eratosthenes, who conceived geography in a basically cartographic and mathematical sense. The most famous ancient advocate of this conception was undoubtedly Ptolemy. In his Geographical Guide,[4] he drew a sharp distinction between geography and chorography. Geography is defined as "the representation, by a map, of the portion of the earth known to us, together with its general features." The task of geography is to present the known world as "one and continuous," so that it may be perceived "as a whole." It is "concerned with quantitative rather than with qualitative matters." The method of geography is "the mathematical"; its calculations must be based on astronomy and geometry. Thus, as for Eratosthenes, geography is defined in a basically cartographic and mathematical sense. On the other hand, chorography, which aims at a description of particular regions, each represented separately,

[2]All remarks and footnotes in the Physische Geographie were added by Rink; see his "Vorrede des Herausgebers," GS, IX, 154.

[3]"The Concept of Geography as a Science of Space, from Kant and Humboldt to Hettner," Annals, AAG, XLVIII, No. 2 (1958), 97-108; quote from p. 99.

[4]For a translation of the relevant passages from Ptolemy, see A Source Book in Greek Science, ed. by Morris R. Cohen and I. E. Drabkin, pp. 162-164.

"does not require the mathematical method." For Ptolemy, geography and chorography are distinct sciences; although they are related in that both are "cartographic," i.e., depict the known world, the method of each is quite distinct.

The second great tradition of geographical definition, in its most comprehensive classical form, stems from Strabo. Although he never specifically defines geography in a short sentence or two, it is evident that his basic conception of the field is that its task is to "describe what is in this our own inhabited world."[5] Geography is a "description of the earth" (6, 1, 2), understood in the sense of its ecumene or inhabited portion. Thus, geography is basically concerned with the earth as "the dwelling-place of man" (1, 1, 16), yet with both its "physical and ethnic" features (4, 1, 1). Nevertheless, Strabo points out that "most of all . . . we need . . . geometry and astronomy for a subject like geography" (1, 1, 20). In addition, the geographer requires a knowledge of "terrestrial history," of animals and plants, and "everything useful or harmful that is produced by land or sea." However, Strabo immediately makes it quite clear that these studies are not geography proper, but "preliminary helps" (1, 1, 16), background knowledge necessary for the attainment of a complete understanding of the earth as the "dwelling-place of man." As we have seen, the geographer must take the findings of physics and astronomy on faith. Human affairs are the measure of ultimate significance in geography, not mathematics or astronomy. Because of geography's practical importance in governing the state, Strabo, in keeping with his Stoic philosophy, sees it as more akin to ethics and politics than to astronomy and mathematics (1, 1, 18).[6]

Standing much closer to Kant in time, Varenius conceived geography in a way that encompassed both the classical definitions, although his basic conception of the field is much more akin to that of Ptolemy. Varenius defined geography as a "branch of mixed [applied] mathematics":

La Géographie est une partie des Mathématiques mixtes, où on explique l'état de laxterre et de ses parties, qui regarde la quantité, sçavoir sa figure, sa position, sa grandeur et son mouvement, avec les apparences célestes, etc.

---

[5]The Geography of Strabo, 2, 5, 13; 2, 5, 34.

[6]Strabo's concept of geography was revived in the seventeenth century by the German geographer, Cluverius.

Il y a des Auteurs qui prenant le mot Géographie dans un sens trop resserré, la définissent une description nue des différens pays. D'autres lui donnant trop d'étendue, voudroient y comprendra la constitution politique des Etats. Ceux qui en usent ainsi, sont excusables en ce qu'ils ne le font que pour encourager et amuser agréablement le Lecteur. En effet il n'apporteroit qu'une foible attention à une énumération et à une description simple des pays, si on n'y joignoit quelque connoissance des moeurs et des coutumes de leurs habitans.[7]

J. N. L. Baker[8] has pointed out that the French version of Varenius, because of several translations and editings, is in places corrupt. Basing his interpretation on the original Latin edition, Baker argues that Varenius did not include human geography as a part of the subject "under protest," as the French version seems to imply, but "because of convention and usefulness" (p. 58). Nevertheless, although Varenius drew a distinction between "general" and "special" geography, and included the study of human particulars under the second division, and although he wrote a work on "special" geography, Descriptio Regui Japoniae et Siam, the latter, except insofar as it included "celestial" demonstrations, does not follow from the basic definition of geography as a "branch of mixed mathematics."

In proving Geographical propositions we are to observe that several properties, and chiefly the celestial, are confirmed by proper demonstrations. But in Special Geography (excepting the celestial) almost everything is explained without demonstrations, being either grounded on experience or observation, or on the testimony of our senses: nor can they be proved by any other means. For science is taken either for that knowledge which is founded on things highly probable; or for a certain knowledge of things which is gained by the force of argument, or the testimony of sense; or for that knowledge which arises from the demonstration in a strict sense, such as is found in Geometry, Arithmetic, and other Mathematical Sciences; excepting Chronology and Geography; to both of which the name of Science, taken in the second sense, doth most properly belong.[9]

Although the meaning of this passage is not altogether clear, it appears fairly evident that "general" geography is to be understood as a science in the third sense, since most of its prop-

[7]Bernhard Varenius, Géographie Générale, revue par Isaac Newton, p. 2.

[8]"The Geography of Bernhard Varenius," The Institute of British Geographers, Transactions and Papers, Publication No. 21 (1955), 51-60.

[9]Trans. by Baker, ibid., p. 59.

ositions are proved by mathematical demonstrations or by
astronomical laws, while "special" geography, which is akin
to chronology, is a science in the second sense. Thus, "general" and "special" geography are distinct sciences, falling
within separate categories; "special" geography is not simply
an application of the principles and rules of "general" geography, nor is it deducible from the statements of "general"
geography, nor can it qualify as geography defined as a branch
of mathematics. Varenius' main objective was to clarify the
principles of "general" geography, and he objected quite strongly to the state of the subject in his day, pointing out that it was
"treated almost exclusively [as] Special Geography, and at te-
dious length."[10] Geography could hardly claim to be a science,
since geographers were ignorant of the mathematical and astronomical foundations of the subject.[11] Nevertheless, in the
dedication to the Descriptio, Varenius made a strong plea for
"special" geography on the grounds of its usefulness to the
statesman. As Baker has rightly observed, in this respect
Varenius stands more in the tradition of Strabo's conception
of geography (p. 55).

Nathaniel Carpenter, perhaps the foremost English geographer of his day, in the second edition of his Geographie delineated forth in two Bookes (Oxford, 1635), had defined geography
simply as "a science which teacheth the description of the whole
Earth." It is a part of "Cosmographie," although Carpenter
carefully distinguishes it from "Astronomie," the other or celestial part of that general science. Geography is divided into
"sphericall" and "topicall" parts, and these parts are respectively dealt with "Mathematically" and "Historically,"[12] although the word "historical," being akin to the Greek word
"historia," had a much wider denotation than it has today.
Again, in Carpenter, we can see the uncompromising division
of the subject into two distinct points of view and approaches.

[10]Quoted in ibid., p. 56.

[11]Varenius was the first geographer to have adequately incorporated the work of
Copernicus, Kepler, and Galileo into his conception of the subject. It was undoubtedly
this aspect of Varenius' Geographia Generalis that so attracted Isaac Newton, who
edited the work for the use of his students.

[12]See A. Wolf, A History of Science, Technology and Philosophy in the 16th and 17th
Centuries, II, 388-390; and R. E. Dickinson and O. J. R. Howarth, The Making of
Geography, p. 139.

This uncompromising division of geography into two kinds of knowledge, which we have already seen in D'Alembert's classification of the sciences, reflects the then prevalent bifurcation of philosophy into rationalism and empiricism. "General" or "sphericall" geography was a rational science, whereas "special" or "topicall" geography was an empirical science.

Kant's basic concept of geography, "a description of the whole earth," where "earth" is understood to mean the surface of the earth, stands much closer to the tradition of Strabo than to that of Eratosthenes and Ptolemy.[13] Not that Kant underestimated the importance of mathematics in geography. As he remarks:

> Before we proceed to the treatise on physical geography proper, we must have a preconception of mathematical geography, as we have already remarked, since we will need it in our treatise only too often. Therefore, we mention here the shape, size and motion of the earth, as well as its relationship to the rest of the world-structure.[14]

Thus, Kant's geography contains a lengthy "Mathematische Vorbegriffe" (pp. 166-183). Kant implies that the geographer should make use of mathematics wherever possible. Although the possibilities for employing mathematics extensively in geography, other than in the usual astronomical contexts, were decidedly limited in Kant's day, his Physische Geographie contains simple mathematical calculations of distances, depths of oceans, heights of mountains, temperature, etc. However, the main point is that Kant did not conceive or define geography as a branch of mathematics. Mathematics is an indispensable tool in geography, but the latter is not a mathematical discipline. Kant defined geography as the "description of the whole earth," and thought of it as a propaedeutic to both science and life. In his basic definition of geography, there was nothing

---

[13] Although to my knowledge Kant never makes reference to Strabo, it is difficult to imagine that he was not aware of the latter's Geography, since the first translation of that work into a language other than Latin was made at Königsberg by Abraham J. Penzel and was published at Lemgo in 1775 and 1777. Furthermore, the dedication of the third and fourth volumes (1777), to the Right Honourable Mr. Ritter, was written in the office of Kanter's book store in Königsberg. Kant boarded with Kanter at the book store for several years during the late 1760's, although by 1777 he had been gone for a number of years. Nevertheless, Kanter's book store was one of the intellectual centres of Königsberg and Kant continued to be a frequent visitor.

[14] Kant, "Einleitung 6," Physische Geographie, GS, IX, 165.

unusual or original. As Rink observes in his remark at the end
of "Einleitung 3": "We are speaking here of knowledge of the
world, and therefore of a description of the whole earth. The
name geography is therefore taken in no other than the ordinary
meaning" (p. 159).

With respect to the relations, similarities, and differences
between geography and history, there is, in Western thought,
an extensive literature. Already at the birth of "professional"
history, with Herodotus, we find a skilful use of geographical
material. However, he never discussed the relations between
the two disciplines, and appears to have regarded them as one
and inseparable, since his work passes almost imperceptibly
from one to the other.[15]

Polybius discussed the place of geography in history, and
explained to his readers why he had not included as much geographical information in his history as previous historians
had in theirs.[16] He regarded geography as a "branch of history," i.e., "historia." Although it is needed to help explain
history proper, Polybius was unwilling to interrupt perpetually
his historical narrative by introducing geographical material.
As he goes on to explain, geography should not be dealt with
in a "disjointed and incidental manner," but should be fully
treated in its own right. Thus, although the two disciplines
are closely related, in that both are parts of "History,"
Polybius indicates that their approaches to the study of "History" are different. Unfortunately, the sections of his history
devoted exclusively to geography have been lost.

Strabo had been an historian most of his life, and turned
to the writing of geography only in old age. As one might expect, his geography abounds in historical material and references. Yet, he never explicitly discusses the relations between
the two disciplines. Perhaps the closest he comes is in the
following short passage: "However, if anything in ancient history escapes me, I must leave it unmentioned, for the task of
geography does not lie in that field, and I must speak only of

[15]Van Paassen sees in Herodotus the first implicit acknowledgment of Kant's dictum
that "Die Historie ist also von der Geographic nur in Ansehung des Raumes und der
Zeit verschieden" (history differs from geography only in respect to space and time),
The Classical Tradition, p. 233.

[16]The History of Polybius (Bk. 3, chaps. 57-59), in Greek Historical Thought,
trans. by A. J. Toynbee, pp. 174-176.

things as they now are. "[17] Geography then deals with the present, history with the past. But geography needs history insofar as it informs the present.

Moving to the century or two before Kant, we have already seen that Francis Bacon held an implicit distinction between geography and history, although within a traditional framework that regarded history primarily as "historia," a factual inquiry into nature. During Bacon's lifetime, and in the following period, one can find many statements linking geography and history, although almost all are part of a tradition that regarded geography as the "hand-maiden of history."[18] For instance, Richard Hakluyt writes of "Geographie and Chronologie (which I may call the Sunne and the Moone, the right eye and the left of all history)."[19] The statement, however, was not original with Hakluyt, but a paraphrase of a passage from David Chytraeus' Variorum in Europa itinerum Deliciae, published in 1594.[20] A few years later, Purchas writes that: "History without that so much neglected study of Geography is sick of a half dead palsy."[21] In 1621, in his Microcosmus, we find Peter Heylyn writing that: "Geographie without Historie hath life and motion but at randome, and unstable; Historie without Geographie like a dead carkasse hath neither life nor motion at all."[22] Later in life, in his Cosmographia of 1649, Heylyn writes:

---

[17] The Geography of Strabo, 12, 8, 7.

[18] The idea that geography during this period was little more than the "hand-maiden of history" has perhaps been somewhat overdone. This is because the Greek word "historia" has often been translated too literally to mean "history" in our contemporary sense of that word. The contemporary sense of the word "history" does not begin to emerge until the eighteenth century. In fact, Kant was probably one of the first to have pointed out that the word "historia" was semantically confusing, since it encompassed two quite distinct ideas—spatial distribution and changes through time. At least one denotation of the term "historia," however, survives today—a "natural history," as the term is employed in biology, often contains no reference to changes through time, but contents itself with describing the present state of biological phenomena in a certain portion of the earth's surface.

[19] The Principal Navigations, Voyages, Traffiques and Discoveries of the English Nation, I, 19.

[20] See E. G. R. Taylor, Late Tudor and Early Stuart Geography, p. 30.

[21] Quoted in ibid., p. 56.

[22] Quoted in H. C. Darby, "The Relations of Geography and History," Geography in the Twentieth Century, ed. by Griffith Taylor, p. 640.

History and Geography, like the two fires or meteors which Philosophers call Castor and Pollux, if joined together crown our reading with delight and profit, if parted, threaten both with a certain shipwreck: and are like two sisters dearly loving, who are not without pity (I had almost said impiety) to be kept asunder. [23]

Heylyn is interesting, however, since he appears to place history and geography on the same footing, instead of subordinating the latter to "history." With increasing knowledge of geography, and increasing ability to handle geographical material, the tradition that regarded geography as the "hand-maiden of history" gradually fades. Yet, as late as the early nineteenth century, one finds the following remark in a major work, J. Pinkerton's Modern Geography (London, 1806):

Geography has been styled one of the eyes of history, a subservience to which study is undoubtedly one of its grand objects; but it would, at the same time, be foreign to its nature to render it a vehicle of history. The proper and peculiar subjects of geographical science are so ample, and often attended with such difficult research, that it becomes equally rash and unnecessary to wander out of its appropriated domain. [24]

This traditional interpretation of the relations between geography and history was also evident on the continent. Even the great Varenius mentions geography's "value to history, whose beacons are rightly said to be chronology and geography."[25] In the following century, Vico discussed geography as "the other eye of history,"[26] the first eye, of course, being chronology. We have already mentioned that D'Alembert regarded geography, at least in part, as one of the two major "offshoots and supports of History."

Thus, there is a long tradition linking geography to history. However, in the centuries immediately preceding Kant, phrases such as "geography is the other eye of history," or "the right eye of history," or "one of its beacons" had become undiscussed clichés, handed down and repeated almost verbatim from generation to generation. Kant appears to have been the

---

[23]Quoted in E. G. R. Taylor, Late Tudor and Early Stuart Geography, p. 141.

[24]Quoted in J. K. Wright, "Some British 'Grandfathers' of American Geography," in Geographical Essays in Memory of Alan G. Ogilivie, ed. by R. Miller and J. Wreford Watson, p. 164.

[25]Quoted in Baker, "The Geography of Bernhard Varenius," p. 55.

[26]The New Science of Giambattista Vico, p. 234.

first to have analyzed the relations, similarities, and differences between geography and history in a way approaching a "modern" perspective.

The other major area in which Kant conceivably made an original contribution to geographic thought concerns the place of geography in a classification of the sciences, and its differentiation from the so-called "systematic" sciences, from biology, and from history. The question of the place of geography in Kant's classification of the sciences is controversial to say the least, and will have to be examined in detail later. Kant, of course, was not the first philosopher to have included geography in a classification of the sciences. A place for that discipline can be found in the classifications of Francis Bacon, Hobbes, and D'Alembert. Among geographers, Strabo, in many scattered references, had hinted at the distinction of geography from physics, astronomy, mathematics, natural history, history, ethics, and politics, but the issue is never really discussed. In fact, such discussion appears to be generally lacking in works by geographers prior to Kant.

In the introduction to the Physische Geographie,[27] Kant draws a distinction between classifying knowledge according to "concepts" (the logical classification), and according to the time and space in which things are actually to be found (the physical classification). The former gives rise to a "system of nature," such as that of Linnaeus, the latter to a geographical description of nature, although geography is also a "system of nature" in some sense of that term. Linnaeus' system of nature, however, was "artificial," a classification we "make in our heads." Kant's examples of the distinction are biological. For instance, the biologist classifies crocodiles and lizards together since they are fundamentally one and the same animal, yet they are found in different places on the earth's surface. There is, of course, nothing unusual in Western thought in geographical classifications of biological phenomena. In fact, the earliest biological classification on record, to be found in the Hippocratic work, On Diet, is primarily geographical. Animals are divided into domesticated and wild, birds into land and water species, and fish into haunters of the shore, free-swimming forms, cartilaginous, mud-loving forms, and

[27]"Einleitung 4," GS, IX, 159-160.

fresh-water species.[28] However, a biogeographical classification of species, especially of marine animals, is often biologically sound because "an ecological classification has its justification, independent of morphological taxonomy, in view of the obvious structural adaptations involved."[29] And Aristotle, of course, was quite familiar with questions of geographical distribution and differentiation of animals.[30] Moving closer to Kant in time, it is undoubtedly true that biologists, from the Renaissance on, possessed a growing awareness of problems of biogeography.[31] And Linnaeus, in his Systema Naturae, often included, in his description of plants and animals, mention of the places on the earth's surface where they were to be found. Whether anyone earlier than Kant had made the distinction as explicitly as he, it was undoubtedly at least implicit in the literature. However, Kant is simply using biological examples to illustrate his main point. He implies that the whole range of natural earth phenomena can be classified either logically or physically, i.e., geographically.

Kant's position in this matter is to be understood against the background of the great debate over the "system of nature" in his day, as it centred principally around the persons of Linnaeus and Buffon.[32] To determine precisely Kant's position respecting this debate would be a vast undertaking. Nevertheless, something of an introductory nature can be said. Despite the opposition in their views, Kant often makes reference,

[28]See Charles Singer, "Biology before Aristotle," in The Legacy of Greece, ed. by Richard Livingstone, p. 169.

[29]Richard Hesse, Ecological Animal Geography, rev. by W. C. Allee, and Karl P. Schmidt, p. 41.

[30]Historia Animalium. For instance, see 589a10 ff., where he begins with the remark that "animals are also differentiated locally"; 602a15 ff., where he begins by remarking, "Particular places suit particular fishes"; and 605b22 ff., where he discusses "variety in animal life as produced by variety in locality."

[31]See Charles E. Raven, Natural Religion and Christian Theology, First Series, Science and Religion, chap. v, "Gesner and the Age of Transition," pp. 80-98.

[32]For recent discussion of Buffon in relation to geography, see Clarence J. Glacken, "Count Buffon on Cultural Changes of the Physical Environment," Annals, AAG, L, No. 1 (1960), 1-21; for discussion of Linnaeus, see C. E. Raven, Natural Religion, chap. viii, "Linnaeus and the Coming of System," pp. 145-163. For a more philosophical discussion of eighteenth-century biology, that includes mention of Kant, Linnaeus, and Buffon, see Arthur O. Lovejoy, The Great Chain of Being, chap. viii, "The Chain of Being and Some Aspects of Eighteenth-Century Biology," pp. 227-241; and Ernst Cassirer, The Problem of Knowledge, Part II, chap. vi, "The Problem of Classifying and Systematizing Natural Forms," pp. 118-136.

usually favourable, to both throughout his works and appears
to have trod a middle road between them. Kant shared with
Linnaeus a "passion for architectonic," an urge to organize,
classify, and systematize knowledge. Linnaeus' Systema Naturae, in fact, ran the whole gamut of nature, from the universe
itself, through men, animals, and plants, to rocks and minerals. Linnaeus even classified diseases. Kant, however,
found the Systema Naturae "artificial"; it was really an "aggregate of nature" and not a "system," since it lacked "the idea
of a whole out of which the manifold character of things is to
be derived."[33] In short, Linnaeus' "system" was philosophically ungrounded. Linnaeus himself had admitted that his
classification of plants and animals was "arbitrary," and not
a true and complete "natural" classification, since it was based
on only one physical characteristic, that of sex. For Kant,
such an artificial ordering of phenomena could not be treated
as scientific knowledge. Cassirer sees Kant, in the Critique
of Judgement, playing the role of "logician to Linnaeus' descriptive science," and attempting to provide it with a suitable
philosophical foundation.[34] On the other hand, Kant could
hardly have accepted Buffon's radical nominalism that regarded
concepts like "species" and "genera" as "artificial," as mere
figments of man's imagination. For the early Buffon, "in
reality individuals alone exist in nature."[35] Nor could Kant
have accepted the evolutionary implications of Buffon's teaching. This was partly because he believed, with Linnaeus, in
the fixity of species, but also because his healthy respect for
empirical evidence led him eventually to reject the "history of
nature" as scientifically desirable but virtually impossible of
attainment. In addition, there were undoubtedly underlying
epistemological problems for Kant, who grounded his phenom-
enalism of nature on space conceived as the sole form of
outer sense. Thus, Kant's epistemology could deal satisfactorily with spatial relations, with a geography of nature, but
not with temporal relations, with a history of nature, since
temporal sequences in nature are not sequences in one's own
experience, and hence cannot come under the form of time.

[33]"Physische Geographie," GS, IX, 160.
[34]Cassirer, The Problem of Knowledge, p. 127.
[35]From Histoire Naturelle, I; quoted in Lovejoy, The Great Chain of Being, p. 230.

Kant could give us an anthropology of human nature, and a geography of "outer" nature, but he could give us only a history of nature conceived as "a continuous geography."[36]

I think what Kant found attractive in Buffon was the panoramic view of nature, the great compendium of nature, conceived in true Baconian fashion. Despite its tendency towards superficiality, the Histoire Naturelle is almost without parallel in Western thought as an attempt to present the entire range of natural knowledge in more or less organized fashion. Kant is attempting something similar in his Physical Geography, although in much more limited and modest fashion. Certainly, for Kant, one of the aims of geography was to present a framework within which all our empirical knowledge of the natural world could find a place. Adickes states that "Buffon is more Kant's man than Linnaeus." In support of this contention he quotes the following passage from one of the manuscript copies of Kant's lectures on physical geography:

The only work in which natural history is properly handled is Buffon's "Epoques de la Nature." However, Buffon gave free rein to his imagination and therefore has written more a romance of nature than a true history of nature.[37]

This is a rather back-handed compliment, but it does illustrate Kant's respect for Buffon, and also his skepticism respecting the possibility of a "true" history of nature. Evidently there is philosophical controversy surrounding the respective importance of both Linnaeus and Buffon as stimulators of Kant's thought.

One of the best illustrations of the middle road that Kant trod between Linnaeus and Buffon is to be found in his theory of race.[38] The theory of race for Kant was an integral part geography, and not of anthropology as he understood that term.[39] Linnaeus had classified races on the basis of a motley

[36]"Einleitung 4," Physische Geographie, GS, IX, 161.

[37]Adickes, Kant als Naturforscher, II, pp. 394n-395n; quoted from p. 6 of Barth's manuscript of Kant's lectures on physical geography, summer session, 1783.

[38]See Walter Scheidt, "The Concept of Race in Anthropology and the Divisions into Human Races from Linneus to Deniker," in This is Race, ed. by Earl W. Count, pp. 354-391; for specific discussion of Kant, see pp. 372-374.

[39]Kant's earliest paper on the theory of race, "On the Different Races of Man," GS, II, 427-443 (trans. by Count, This is Race, pp. 16-24), was written as an announcement of his lectures on physical geography for 1775.

array of characteristics: physical, cultural, and "temperamental" understood along traditional Hippocratic lines. Different races, however, were original and distinct. Buffon had classified races geographically, using principally physical characteristics such as skin-colour, stature, and bodily figure as indices. He postulated an original homogeneous species of man that became differentiated geographically on the basis of environmental factors, principally climate. Kant accepts the geographical differentiation of the races—white, yellow Indians, Negroes, and copper-red Americans—distinguished principally on the basis of skin-colour.[40] Although the races may be differentiated geographically, the reasons for the differences between them are not geographical, i.e., cannot be explained on the basis of climate, etc., but lie in the original endowment, in the "germs" (Anlagen) of the separate races.[41]

THE DEVELOPMENT OF KANT'S
CONCEPT OF GEOGRAPHY

This section is not a detailed study or assessment of all Kant's works that might be considered geographical, either in whole or in part. It is concerned more with his concept of the subject than with specific content and the main line of development is fairly clear.

Kant's earliest statement of his concept of geography occurs in his Entwurf und Ankündigung eines Collegii der physischen Geographie, of 1757.[42] Kant points out that the study or consideration of the earth is threefold. The mathematician sees only the form of the earth and views it as devoid of creatures. "Political doctrine" teaches us to know people and their communities, through the forms of government, commerce, mutual interests, religion, and customs. Kant gives no clear indication here that these ways of studying the earth are to be regarded as parts of geography, although later he employs the

[40]"Bestimmung des Begriffs einer Menschenrace" (1785), GS, VIII, 93.

[41]Ibid., p. 98. In this connection, it is interesting to observe that a leading authority on the history of the theory of race has recently written: "Apparently it is now almost forgotten in the United States that Immanuel Kant produced the most profound raciological thought of the eighteenth century" (Count, This is Race, p. 704).

[42]GS, II, 3-10.

term "mathematical geography" (p. 4) to cover the mathematical view of the earth. However, he does indicate that these approaches were being adequately handled in his day. It was otherwise with physical geography, the third way of considering the earth.

Physical geography considers only the natural condition of the earth and what is contained on it: seas, continents, mountains, rivers, the atmosphere, man, animals, plants and minerals. All this, however, not with the completeness and philosophical exactitude which is the business of physics and natural history, but with the reasonable curiosity for the new of a traveler who seeks out everywhere what is noteworthy, peculiar and beautiful, and compares his accumulated observations according to some plan [p. 3].

Evidently, geography is not a science in the sense in which physics and even natural history are sciences. Nevertheless, it is not to be regarded as a mere aggregate of information, for Kant goes on to point out that he had made the information collected from various sources into a "system" (p. 4). Even so early in his thoughts on the subject, geography is to be regarded as a system in some sense of that term. The passage also indicates that geography is a study of the present, based on careful observation, and by implication, of phenomena on or near the earth's surface. Yet, in the brief outline of the content of his course, he presents the various aspects of geography as history (Geschichte). This is also borne out by sections 53 following of Rink's edition of the Physische Geographie, which in Adickes'[43] opinion date from before 1760. Here, the various parts of geography are presented as "Geschichte." Apparently at this time Kant had not made any clear distinction between geography and history, and geography is presented as a sort of popularized natural history.

The geographical view of man at this stage is also of interest. Geography evidently does not study the whole of human activity, since apparently "political doctrine" also studies man. Yet, man is included in physical geography. The Physische Geographie contains a section on man (pp. 311-320), in which he is regarded as a part and product of nature, viewed from the perspective of the natural environment. Tatham has translated representative passages as follows:

[43] An outline of Adickes' findings respecting Kant's course of lectures on geography will be given later in this chapter.

All inhabitants of hot lands are exceptionally lazy; they are also timid and the same two traits characterize also folk living in the far north. Timidity engenders superstition and in lands ruled by kings, leads to slavery. Ostoyaks, Samoyeds, Lapps, Greenlanders, etc. resemble people of hot lands in their timidity, laziness, superstition, and desire for strong drink, but lack the jealousy characteristic of the latter since their climate does not stimulate their passions so greatly.

Too little and also too much perspiration makes the blood thick and viscous. . . . In mountain lands men are persevering, merry, brave, lovers of freedom and of their country. Animals and men which migrate to another country are gradually changed by their environment. . . . The northern folk who moved southward to Spain have left a progeny neither so big nor so strong as they, and which is also dissimilar to Norwegians and Danes in temperament. [44]

As Tatham points out: "There is nothing either fresh or particularly geographical in these remarks. They are almost identical in form and content with those of Montesquieu." Nevertheless, physical geography aims at more than just presenting man as a product of the natural environment. Kant remarks in his outline:

I present this material first in the natural order of classes, and proceed later to set forth, in a geographical manner of instruction, through referring to all countries of the earth, the particular inclinations of men which derive from the regions in which they live, the variety of their prejudices and ways of thinking, in so far as all this serves to acquaint man better with himself, including a brief comprehension of his art, commerce and science, and a report on the products of the above mentioned lands in their proper places, climate, etc., in a word, everything that belongs to a physical consideration of the earth [p. 9].

The term "physical" is employed here in a very broad sense, and evidently includes much that pertains to what "acquaint[s] man better with himself."[45] In this passage, Kant also draws a distinction between considering geographical material in terms of various classes, i.e., a "systematic" or subject-matter approach, and considering the material regionally; as it is found in association in various countries throughout the world. In this latter context, man is the central focus. But,

[44]Tatham, "Environmentalism and Possibilism," in Geography in the Twentieth Century, ed. by Griffith Taylor, pp. 128-162; quote from pp. 130-131.

[45]See Schilpp's interesting comment on this passage, to the effect that Kant's aim of making man "'more intimately acquainted with himself' reveals a most interesting [early] hint of reflective self-criticism" (Paul A. Schilpp, Kant's Pre-Critical Ethics, p. 20).

in the early edition of the Physische Geographie, some of the
material in the non-regional sections is also related to man,
and the usefulness to man of the information is often mentioned.[46] Thus, one of the early aims of Kant's geography was
clearly to present useful information that might enable man to
understand himself and to live better.

In the second major outline of his courses,[47] Kant introduces the section on physical geography (pp. 312-313) by remarking that he had started at the very beginning of his teaching
career to present geography in the form of a sum (Inbegriff) of
knowledge that was agreeable and easily understood. It was
necessary, in Kant's opinion, particularly that young students
who lacked experience of the world be given a general framework for knowledge within which to orient themselves. Thus,
the importance of geography in education was apparent to Kant
at a very early stage, even before he was influenced by
Rousseau, and geography's role as a propaedeutic begins to
emerge. Kant goes on to point out that he had gradually expanded his original outline, and he now proposes a physical,
moral, and political geography. This "new" subject was apparently to be regarded as an integrated discipline and not just
as three distinct fields of knowledge. Yet, physical geography
proper continues to study the natural interconnections between
land and sea, and is the true ground of all history, without
which the latter is little more than a collection of fairy-tales.
Moral geography studies what is moral in man, the variety of
and differences between his natural attributes, over the whole
earth. This comparative study, both historical and contemporary, will eventually produce a sort of "moral map" of the
human race. The study of political geography aims not so
much at an understanding of the transient features of states,
such as the succession of their rulers, but at an understanding
of more permanent features such as the position and situation
of various countries, their produce, customs, trade and com-

[46]For instance, man is depicted as one of the principal agents bringing about changes
in the face of the earth (p. 298); the more a seafarer knows about coastlines, depths of
oceans, causes of winds and storms, etc., the better sailor he will be (pp. 306-307);
various plants are presented in terms of their usefulness to man (pp. 356-362); wine
making is discussed (pp. 364-365).

[47]"Nachricht von der Einrichtung seiner Vorlesungen in dem Winterhalbenjahre von
1765-1766," GS, II, 303-313.

merce, and population. In general, the revival of the science of geography should bring about "that unity of knowledge, without which all learning is only piece-meal (Stückwerk)." A remarkable and important role for geography, to say the least! Evidence of this much wider horizon for geography, however, comes even earlier than 1765, since Herder's lecture notes of the winter session 1763-64 have been preserved. During that session, Kant presented physical geography as "(a) a teacher of morality, which reveals the unaffected savage (ungekünstelten Wilden); (b) an instrument for political geography; (c) a summary of natural history and the key to theoretical physics; (d) also to theology, etc.; (e) it also brings ennobling satisfaction."[48]

The mid-1760's mark the years of Kant's greatest flirtation with empiricism. As might be expected, and as indicated in the striking passage quoted above, geography, as the empirical study of nature conceived in a broad sense, occupies a place of central importance in Kant's thinking during this period. In fact, never again was geography to occupy a position of such importance, nor to be understood in so broad a sense. Although one can find scattered geographical references and content in writings of this period, the work that is most characteristically geographical is one that a recent commentator has described as "the epitome of Kant's pre-Critical thought," Observations on the Feeling of the Beautiful and Sublime.[49] It has been customary to regard this book as a work in aesthetics, and as a forerunner to the aesthetic doctrines of the Critique of Judgement. However, given Kant's very broad concept of geography of the time, an equally strong case could be made for regarding it, at least in part, as a geographical work. Kant's approach is empirical and inductive, based on keen observation of nature and of the human scene. Beauty is objective; it resides in nature. Thus, in places, Kant presents us with a sort of aesthetic geography. But especially section 4, "Of National Characteristics," can be construed as moral and political geography, although Kant does point out: "Whether

---

[48]Quoted in Adickes, Kant als Naturforscher, II, p. 380. Adickes goes on to remark that (a) was certainly conceived under Rousseau's influence, and probably also (b) and (d).

[49]Trans. by John T. Goldthwait, p. 1.

these national differences are contingent and depend upon the times and the type of government, or are bound by a certain necessity to the climate, I do not here inquire" (p. 97). Kant, of course, lectured for many years on empirical aesthetics and national characteristics in his course on anthropology. [50] And so, the Observations should perhaps be regarded more as a forerunner to the Anthropology than to the Critique of Judgement. However, at this time, 1764, Kant had not yet separated anthropology from geography, and much of what later became "moral" anthropology is included under the general term "moral and political geography, " which, as we have seen, studies human customs and "what is moral in man, the variety of and differences between his natural attributes, over the whole earth." The term "natural attributes" appears to suggest some limitation for geography, and Rousseau's concept of the "natural man" or the "noble savage" comes strongly to mind. Nevertheless, national characteristics are discussed in the Observations in terms of the traditional fourfold Hippocratic classification of temperaments—melancholy, sanguine, choleric, and phlegmatic. In a different context, Kant indicates that when psychologically considered, temperament can only be treated by analogy since it is "in the blood, " as a physiological or natural endowment. [51] It does not follow, of course, that the possession of a certain temperament is going to determine the whole of one's character. However, Kant says too little on moral geography to admit of a clear solution to the question of the limitations he set for such a study at this time. Although an abstract of moral geography continued to be given in the introduction to the Physische Geographie even after he separated anthropology from it, there is no evidence that he ever gave a course on "moral geography. "

The later relations between "moral" anthropology, moral geography, and ethics are best set forth in Kant's Lectures on Ethics. Kant points out that "anthropology observes the actual behaviour of human beings and formulates the practical and subjective rules which that behaviour obeys, whereas moral

---

[50]"Anthropologie, " GS, VII, 230-250, 313-320. In addition, in Bk. III of Part I, pp. 251-282, which might be described as "moral" anthropology, Kant often mentions differences in the customs of various peoples.

[51]Ibid., p. 286.

philosophy alone seeks to formulate rules of right conduct, that is, what ought to happen" (p. 2). These two sciences, practical philosophy and anthropology, are closely related:

> . . . we can pursue the study of practical philosophy without anthropology, that is, without the knowledge of the subject. But our philosophy is then merely speculative, and an Idea. We therefore have to make at least some study of man [pp. 2-3].

Moral geography should find a place in this "study of man," but where? Since "anthropology is a science of the subjective laws of the free will, [and] practical philosophy is a science of its objective laws" (p. 3), is there any place left for a moral geography? Kant appears to find a place on the basis of his distinction between inner and outer sense. In considering morality empirically, an anthropological or inner sense approach to the study of ethics is based on feeling, a geographical or outer sense approach on custom (or mores). This distinction implies that an anthropological approach deals with self-conscious behaviour, whereas a geographical approach deals with unreflective behaviour. Thus, moral geography may discover that it is customary to permit theft in Africa, to desert children in China, to bury them alive in Brazil, and for Eskimos to strangle them. But this knowledge is of slight importance to ethics, for as Kant remarks, if "we act by reference to customary example and the commands of authority . . . it follows that there is no ethical principle, unless it be one borrowed from experience" (p. 13). Thus, whatever the precise status of moral geography may have been for Kant in the 1760's, evidently by the 1780's it was of little real philosophical importance.

The only other work of Kant's earlier period which is of direct philosophical interest, although admittedly peripheral to the main line of development of his concept of geography, is his paper of 1768, On the First Ground of the Distinction of Regions in Space.[52] What is of interest in this paper is Kant's pronouncement that even if, for instance, we carry a complete chart or plan of the heavens in our minds, and in addition, know where north is, we cannot orient ourselves to this chart,

---

[52]Contained in Kant's Inaugural Dissertation and Early Writings on Space, trans. by John Handyside, pp. 19-29. (The term "region" in Kant's paper bears no direct relation to the technical geographical concept "region"; the term is perhaps better translated as "direction.")

or derive other directions from it, unless we can first orient ourselves to it relative to our own bodies, to our right and left hands. In addition, the direction north itself has no meaning aside from this basic orientation in terms of our right and left hands. Anyone who is used to working with maps will know how true this is. A map that lacks a north pointer is a useless tool; it possesses an air of unreality. But once one has located the north pointer, one orients oneself to the map by automatically associating east with the right hand and west with the left hand. And by this very act of bodily association, the concept north itself takes on meaning relative to the other directions. Kant goes on to point out:

> Similarly, our geographical knowledge, and even our commonest knowledge of the position of places, would be of no aid to us if we could not, by reference to the sides of our bodies, assign to regions the things so ordered and the whole system of mutually relative positions [p. 23].

Kant returned to this theme years later, in a paper of 1786.[53] He draws a distinction between orientation in thinking, or logical orientation, and geographical orientation. But the basic meaning of the term is geographical.

As an example of geographical orientation, Kant says that if I see the sun in the sky and know it is noon, I can find south, west, north, and east on the basis of the distinction between my right and left hand. In short, I orient myself geographically with all the objective facts of the sky only through a subjective distinction. It seems to me, however, that one piece of objective information is initially necessary—knowledge as to whether one lived in the northern or the southern hemisphere, i.e., whether the sun's path crossed the southern or the northern half of one's celestial horizon. Another example given by Kant is interesting. In order to orient myself, find my position, in a completely dark room with which I am familiar, I need grasp only one known object. Yet, it would appear that an initial assumption is necessary here. Assuming that I am suddenly set down in the room, if the first object I grasp is symmetrical, say a round table in the middle of the room, I would know that I was near the centre of the room, but I would still not know my position relative to other objects. In this case, I

[53]"Was heiszt: Sich im Denken orientiren?" GS, VIII, 131-147.

71

would need to grasp a second object. But if the room were symmetrically filled with round tables, I would have serious trouble orienting myself. In short, an underlying assumption, in orienting oneself in the space of the earth's surface, is that that space is asymmetrical with respect to its objects.

The central statement of Kant's mature thought on geography is contained in the introduction to the Physische Geographie. The earliest record of the introduction that Adickes was able to find is in a manuscript copy of Kant's lectures on geography given in the summer session of 1775. It would seem most reasonable, however, to date it from either 1772 or 1773, since Kant separated anthropology from geography and began to lecture on the former subject in the winter session of 1772-73. In the introduction to the Physische Geographie, Kant separates anthropology from geography on the basis of his distinction between inner and outer sense. Undoubtedly, the introduction contains ideas that had been expressed earlier, but it also contains new ideas on geography and on the relations of geography to other disciplines. My assumption, therefore, is that it is to be understood in terms of Kant's more mature thought. Kant apparently never saw any reason for altering this introduction substantially, so that it must have fitted in reasonably well with his later doctrines.

During the early 1770's Kant's thought was undergoing considerable change generally. In fact, Kant's letter of February 21, 1772, to Marcus Herz, in which for the first time he projects the writing of a Critique of Pure Reason, has sometimes been described as the "birth of the critical philosophy."[54] During this period, Kant's thoughts on geography also underwent an upheaval. Some additional light is shed on Kant's concept of geography in certain passages in the Critique of Pure Reason, and in his works on race and on history. In addition, teleological aspects of geography, which are not considered in the introduction to the Physische Geographie, are discussed in the Critique of Judgement (including its first introduction) and in his article of 1788, Über den Gebrauch teleologischer Principien in der Philosophie. And, of course, there is the Anthropologie, which must be considered, since Kant always thought of geography and anthropology as closely related, as forming a "whole" of our empirical knowledge of the world.

[54]See Kant, trans. and ed. by Rabel, pp. 96-97.

72

KANT'S INFLUENCE

The reception of Kant's ideas on geography has had a rather peculiar history. His lectures on geography undoubtedly enjoyed considerable popularity during his lifetime. The course itself was well attended by persons with a wide range of cultural and educational backgrounds. Handwritten copies of the lectures are known to have been circulated widely, and in 1778, we even find the Prussian Minister of Culture, von Zedlitz, a great admirer of Kant, writing to the philosopher to tell him that he had just read, with much pleasure, an imperfect manuscript copy of the lectures, and to request a more correct copy.[55] Herder probably did something to spread and popularize Kant's views on geography. He was greatly influenced by Kant's course of lectures on the subject, and geography was always an important element in and foundation for his own philosophy of history. Nevertheless, not the least criticism in Kant's otherwise mild reviews of Herder's Ideas toward a Philosophy of the History of Man, was directed to his misuse of geographical material, and especially to his romantic and anthropocentric rendering of geographical propositions.[56] There can be little doubt that Kant's authority played an important role in making geography respectable during the early formative years of its modern period. A comparable case in our own age would be Einstein insisting on the importance of geography to all students entering university.

---

[55]Adickes, Kant als Naturforscher, II, pp. 392-393.

[56]Note, for instance, the ironic and somewhat sarcastic way in which Kant calls attention to statements by Herder. "We will leave it to the critics of discriminating philosophic style or to the finishing touch of the author himself to decide, for instance, if it were not better to say: 'not only day and night and the rotation of seasons modify the climate,' than what appears on p. 99: 'Not only day and night and the round dance of alternating seasons modify the climate.' We will leave it to them to decide if the following, doubtlessly beautiful image on p. 100 expressed in a dithyrambic ode is suitably adapted to a natural historical description of these modifications: 'Around the throne of Jupiter the hours (those of the earth) perform a round dance, and whatever is formed under their feet only represents indeed an imperfect perfection because everything is founded on the fusion of heterogeneous elements; but through an inner love and communion with one another is the child of nature—physical regularity and beauty—everywhere born.'" From Kant's "Third Review of Herder," trans. by Robert E. Anchor, in Kant on History, ed. by Lewis W. Beck, pp. 45-46. Clearly, for Kant, Herder's treatment of geography represented a step back.

Following Kant's death, however, the situation respecting his geography becomes obscure and confused. Undoubtedly, new philosophical movements such as romanticism and German idealism on the one hand, and materialism on the other, served for a time to eclipse Kant's philosophy. In addition, the main source of information on Kant's geography was contained in undergraduate lecture notes, hardly the sort of source to which people would go with confidence. Confusion was added by the fact that there were conflicting editions of the Physische Geographie, and consequently distrust of the authenticity of any version. This confused situation was certainly not helped by the Rosenkranz and Schubert edition of Kant's works, which grouped his cosmogony and early scientific papers together under the heading "Physical Geography."[57]

In fact, the confusion and controversy surrounding Kant's geography was not cleared up until the work of Erich Adickes appeared between 1911 and 1913.[58] The controversy begins with the foreword to Kant's Anthropologie,[59] where he announced, with regret, that because of advanced age and the illegibility of his own notes, it would be impossible to produce an edition of his Physische Geographie. However, in 1801, Gottfried Vollmer issued a book purporting to be Kant's geography. This edition was condemned by Kant, and he called on his assistant, Rink, to edit an official version. Despite Kant's condemnation, the Vollmer edition continued to be consulted; for instance, Peschel used it for an inadequate assessment of Kant in his Geschichte der Erdkunde bis auf Alexander von Humboldt und Carl Ritter. The Rink edition appeared in 1802. By this time, Kant was apparently too old and feeble to realize what Rink had done. Adickes, basing his findings on an examination of some twenty manuscript copies of Kant's lectures (dating from as early as the late 1750's to as late as 1792), came to the conclusion that the Physische Geographie, from

[57]See Kant's Cosmogony, trans. and ed. by W. Hastie, p. xviii.

[58]The main source of information is Adickes' Untersuchungen zu Kants physischer Geographie. Brief summaries of the essential points may be found in Kants Ansichten über Geschichte und Bau der Erde, pp. 2-5; Ein neu aufgefundenes Kollegheft nach Kants Vorlesung über physische Geographie, pp. 9-11; and Kant als Naturforscher, II, 373-377. Brief summaries may also be found in Gedan's notes to the "Physische Geographie," GS, IX, 510-513; and in Hartshorne, The Nature of Geography, pp. 38-39.

[59]GS, VII, 121.

74

section 53 to the end in the official Rink edition, dated from
before 1760 and in fact was based on Kant's original notes for
the course, much of which material he had long since abandoned.
On the other hand, the first fifty-two sections, with the exception of sections 11 and 14, were probably compiled from several
later manuscript copies of the lectures. Since Adickes could
find in no other manuscript copy of the lectures material comparable to sections 11 and 14, he came to the conclusion that
Rink had written them, in addition to adding all the remarks and
footnotes. However, the rest of the first fifty-two sections compare favourably, with respect to meaning, to other later copies
of the lectures, and may confidently be accepted as authentic.
Of crucial importance here is the fact that the introduction to
the Physische Geographie, the only really significant part of
the work, can be taken as genuine, and as representing Kant's
more mature thoughts on the nature of geography.

Let us turn to the question of Kant's possible influence on
thought on the nature of geography, especially on Alexander von
Humboldt and Carl Ritter, the men widely looked upon as the
real founders of modern geography. The question of Kant's influence is still an unsettled one and has most recently been discussed by Tatham and by Hartshorne;[60] both indicate that the
evidence is speculative and not conclusive.

Important facts concerning Alexander von Humboldt's life
in relation to Kant are well known. For a time, around 1785,
he was a member of Marcus Herz's Kantian circle in Berlin.
Herz was well acquainted with Kant's lectures on geography.[61]
In addition, Alexander's brother Wilhelm was steeped in Kantian
philosophy during his unversity days, and had apparently read
all Kant's works. Alexander von Humboldt was himself well
acquainted with Kant's writings, but whether he was aware at
this time of the content of the lectures on Physical Geography,
which were not published until 1802, is unknown. In Harts-

[60]Tatham, "Geography in the Nineteenth Century," pp. 42-59; Hartshorne, "The Concept of Geography as a Science of Space, from Kant and Humboldt to Hettner." My consideration of the topic, unless otherwise indicated, is based primarily on their findings,
which are more detailed than my own. At this point, I have only a few footnotes to add
to the debate.

[61]Kant had written to Herz as early as late 1773 respecting his new courses on physical geography and anthropology. See Kant: Philosophical Correspondence, trans. and
ed. by Arnulf Zweig, pp. 76-79, for a translation of this letter.

horne's opinion, it is "almost certain" that Alexander made "use of the statement which Rink published for Kant in 1802" (p. 101), and that his basic concept of the nature of geography accorded well with Kant's. Controversy surrounds the question whether von Humboldt's first statement on the nature of geography, as it appeared in his Florae Fribergensis Specimen,[62] published in 1793, was original or was derived from Kant. That this statement, published nine years before Kant's lectures appeared in the Rink edition, was basic to Humboldt's thought is evident from the fact that he continued to reproduce it from time to time, even in his final work, Kosmos, in substantially unaltered form. However, one important piece of evidence has been overlooked in this matter—Kant's debate with the geographer Georg Forster.[63]

In 1786, in the journal Deutscher Mercur, Forster had raised some objections to a paper by Kant (Bestimmung des Begriffs einer Menschenrace) that had appeared the previous year. Kant replied to Forster in 1788, five years before von Humboldt's statement appeared. The essence of Kant's reply is that Forster disapproved of, and indeed failed to recognize, the distinction between a history of nature and a description of nature. In fact, Kant points out, Forster's own conjectures on the origin of the Negro race clearly belong to a history of nature and not to its description, despite the fact that Forster himself regarded a history of nature, which inquired into origins, as an empirical impossibility, as, of course, did Kant himself. Kant goes on to insist (pp. 162-163), in perhaps the strongest statement he ever wrote on the topic, that it is absolutely necessary, in order that sciences are not allowed to overlap and thus obscure knowledge, that heterogeneous subject matter must clearly be separated. On this basis, a description of nature must be distinguished from a history of nature. Kant then points out that the underlying problem was really a semantic one, since the word "history" was still being used in the Greek sense of "historia," which meant both a tale and a description. Such semantic confusion could be avoided, in Kant's opinion, by using the term "physiography" for the description

[62]For a translation of the relevant section on the nature of geography, see Hartshorne, "The Concept of Geography, " p. 100.

[63]"Über den Gebrauch teleologischer Principien in der Philosophie, " GS, VIII, 157-184.

of nature, and "physiogony" for the history of nature. What
is most striking in von Humboldt's statement of 1793 is the
passage following his brief outline of what geography does, and
how it is distinguished from more formal sciences and from
natural history: "This is what distinguishes physiography from
nature study, falsely called natural history." In view of the
fact that the young Alexander von Humboldt spent a considerable
part of the year 1790 in the company of Georg Forster, who
effectively introduced him to the study of geography, in addition to maintaining a close association with his brother Wilhelm
at this time, it seems highly unlikely to me that he was not
aware of this debate between a philosopher as famous as Kant
and a geographer as famous as Forster, and of their controversy over the distinction between a description of nature and a
history of nature. Evidently, von Humboldt sided with Kant in
this debate, and also drew the distinction sharply. This evidence, of course, is not conclusive, but it does represent the
strongest possible link that could connect the youthful von
Humboldt with statements by Kant.

There is, as Hartshorne has pointed out, evidence of later
direct links between Kant and von Humboldt. In the course of
lectures he gave at the University of Berlin in 1827-28, he indicated that the title, "Physische Weltbeschreibung," had been
taken from Kant. His Kosmos contains statements in quotations,
although without acknowledgement, that bear a striking similarity to statements made by Kant in the introduction to his
Physische Geographie, notably to the distinction between outer
and inner sense. It is on the basis of this evidence that
Hartshorne concludes von Humboldt made use of Kant's introduction. On the other hand, von Humboldt in his Kosmos also
saw fit to attack Kant on occasion. His criticism was directed
mainly to Kant's astronomical theories and other early physical
works, which he described as "divined," "suspected," or
"dreamed." This is a good illustration of the extremes to
which von Humboldt carried the idea of a "description of nature" in opposition to anything that suggested a "history of nature." In this connection, one is reminded of J. T. Merz's
shrewd observation: "The 'Kosmos' of Humboldt closed the
older, the 'Origin of Species' of Darwin opened the new, epoch
of natural science: the former was retrospective, the latter
prospective."[64] In support of this contention, Merz points out

[64]A History of European Thought in the Nineteenth Century, II, 329.

that von Humboldt rarely concerned himself with discussing
genetic theories of nature. For instance, he hardly noticed
Lyell's important Principles of Geology. Merz quotes a number of relevant passages from the Kosmos. The following
short passage will suffice:

The mysterious and unsolved problems of development do not belong to the
empirical region of objective observation, to the description of the developed, the actual state of our planet. The description of the universe,
soberly confined to reality, remains averse to the obscure beginnings of a
history of organic life, not from modesty, but from the nature of its objects
and its limits.[65]

In the year that Alexander von Humboldt died, 1859, Charles
Darwin's Origin of Species appeared. As a result, in a very
real sense, von Humboldt's monumental Kosmos fell still-born
from the press. Darwin's work revolutionized thinking in the
second half of the nineteenth century, and in the revolution,
Kant's and von Humboldt's concept of geography as a "description of nature" was swept aside. In fact, geography itself
tended to disappear in the face of an extended interest in genetic studies. A great period of investigation into historical
geology and geomorphology ensued. On its physical side, geography tended to be absorbed by geophysics and geology, and on
its human side by history and the emerging social sciences.

    In order, however, to get a rounded view of Kant's influence
on von Humboldt, it would be necessary to study also the possible influence of others on the foundations of his thought. His
own brother Wilhelm, although a dedicated Kantian in his youth,
was later influenced by Schiller, Goethe, and Schelling, and
was, in addition, a thinker of no mean originality in his own
right, having done significant work in philology and history.
However, if we are to believe Wilhelm's own assessment of his
intellectual relations with his brother, any possible influence
would have been very slight.[66] Alexander, also, was influenced

---

[65]Ibid., II, p. 277; quoted from Alexander von Humboldt, Kosmos, I, 367.

[66]"Since we were children we [my brother Alexander and I] have diverged like two
opposite poles, although we have always loved one another and at times have confided
in one another. His striving was directed, from an early age, toward the outside
world. And I very early chose for myself an inner life alone. Believe me, in that
difference is encompassed everything." "You know Alexander's views. They can
never be the same as ours, much as I love him. It is often downright funny when he
and I are together. I always let him talk and have his way, for what's the use of con-
tending when the first bases of all our principles are totally different." From letters
to Caroline von Humboldt, March, 1804, November, 1817, contained in Wilhelm von
Humboldt, Humanist without Portfolio, trans. by Marianne Cowan, pp. 386, 407.

for a time by Schelling's philosophy of nature. This influence, however, appears to have been short-lived, since his course of lectures given at Berlin in 1827-28 was specifically directed, in part, against the natural philosophy of Schelling and Hegel.[67] Nevertheless, an aura of romanticism persisted, and can be found in certain mystical and aesthetic statements on nature in general, and in his concept of nature as "one great whole animated by the breath of life."[68] Undoubtedly, however, the greatest single influence on his orientation in thought was the growing scientific empiricism of his age. He was not especially philosophically inclined, and does not appear to have been substantially influenced by any single philosopher other than Kant.

It was otherwise with Carl Ritter, the second great founder of modern geography. He was philosophically inclined, and was evidently much influenced by the philosophies of his day. Although passages can be found in his writings that bear resemblance to statements by Kant, any possible link with Kant is much more tenuous than in the case of Alexander von Humboldt. Ritter was apparently influenced by Herder's philosophy of history, and there may possibly be an indirect link here with Kant. But, as we have seen, Kant was often unimpressed with Herder's use of geographical material, and with his romantic and anthropocentric rendering of geographical statements. The concept of a "Ganzheit," or organic whole, which Ritter may have adopted from Herder, is not the concept of a "whole" that Kant employed in connection with geography. If, as Tatham maintains, Ritter's empirical work was not adversely affected by the teleological principle he employed, then his use of that principle may have been more akin to Kant's, i.e., as regulative of ideas and not constitutive of nature. These are questions which have not as yet been satisfactorily explored.

A wide range of conflicting opinion surrounds interpretation and understanding of Ritter's work. Some have regarded him as purely an historian; others have maintained that his metaphysical and religious views were detrimental to his scientific work; still others have regarded him as an empirical scientist who did significant work in laying the foundations of geography

[67]See L. Kellner, Alexander von Humboldt, pp. 115-117.

[68]From the introduction to Kosmos; quoted in Tatham, "Geography in the Nineteenth Century," p. 53.

as a chorological science. [69] Van Paassen has recently indicated that his thought should be studied in the light of Schelling's natural philosophy. [70] And before that, Hettner, who had had philosophical training, pointed out that Ritter was a "child of his time" and that in the general orientation of his thinking "was under the spell of Schelling, Hegel, and Wilhelm von Humboldt. "[71] There would appear to be a need to study the foundations of Ritter's thought initially in terms of the dominant philosophical ideas of his own period, even before attempting to determine any possible Kantian influence on his thought.

When geography began to re-establish itself as a separate discipline in the last quarter of the nineteenth century, geographers, in their search for adequate theoretical foundations for the subject, turned back to examine the ideas of von Humboldt and Ritter. But there was no return to Kant. In fact, major historians of geography of the period thought of him as a minor eighteenth-century figure. For instance, Oskar Peschel's authoritative Geschichte der Erdkunde bis auf Alexander von Humboldt und Carl Ritter rarely mentions Kant. [72] This tradition regarding Kant was carried on well into the twentieth century by Dickinson and Howarth, [73] and by Leighly. [74]

---

[69] For a recent review of conflicting opinions on Ritter, see de Jong, Chorological Differentiation, pp. 100-101, 131-133.

[70] The Classical Tradition, p. 358. In fact, this has already been done to some extent by van Paassen himself in his article, "Carl Ritter Anno 1959, " Tijdschrift van het Koninklijk Nederlandsch Aardrijkskundig Genootschap, LXXVI, No. 4 (1959), 327-351.

[71] Hettner, Die Geographie, p. 300.

[72] Most references to Kant (pp. 721, 733, 734, 805, 806-807) indicate that his ideas either were not original or were mistaken. The only favourable references are to his theory of the winds (p. 768), and to the fact that, along with Bergman and J. R. Forster, he was one of the first to present a physical geography (p. 808). The introduction to the Physische Geographie is not mentioned.

[73] The Making of Geography. Here we find some of the same minor points mentioned by Peschel (pp. 119, 172). Attention is also drawn to Kant's remarks on the parallelism between the east and west shores of the Atlantic Ocean (p. 172). In addition, there is a very brief résumé, without comment, of the introduction to the Physische Geographie (pp. 119-120).

[74] For Leighly, Kant's importance appears to stem solely from the fact that he was one of the early systematizers of physical geography (John Leighly, "Methodologic Controversy in Nineteenth Century German Geography, " Annals, AAG, XXVIII, No. 4 [1938], 238-258; see especially pp. 239-240).

By the time we reach Hettner, the major theoretical geographer of the late nineteenth and early twentieth centuries, the line of connection with Kant appears to be fairly clear, and is, in fact, rather remote and indirect. In his early paper on the history of geography in the nineteenth century, Hettner does not even mention Kant.[75] Even in the historical introduction to his Die Geographie, Hettner mentions Kant only twice: his General Natural History and Theory of the Heavens is referred to as a forerunner to Laplace (p. 68); the works of Lulofs and Bergman included the main points of Kant's lectures on physical geography (p. 71). In one of his major statements on the nature of geography, Hettner, in a footnote, indicates that he had just learned of similar statements made by Kant from Kaminski's recently published dissertation on Kant's Physische Geographie.[76] This serves as a good illustration of how low knowledge of and regard for Kant's work on the nature of geography had sunk, at this time, when Hettner, who had been trained in philosophy as well as in geography, was apparently unaware of its existence as late as 1905. In his major work, Die Geographie, in discussing the nature and task of geography, and especially its place in relation to other sciences, Hettner quotes a number of passages from the introduction to Kant's Physische Geographie with evident approval, since he remarks that: "Kant had already expressed these thoughts excellently in his lectures on physical geography" (p. 115). In a footnote, he points out that Kant's statements had long "escaped" him, but that he was very glad of the agreement between his own views and those of the great philosopher. One should not assume from these remarks, despite the apparent similarity in ideas, that Hettner based his findings on Kant. The clear indication is that Hettner developed his ideas independently of Kant, but was glad of the confirmation and historical support that the latter offered. Nor did Hettner consider himself to be a Kantian. He points out that the theory of evolution (Deszendenztheorie) stood between him and Kant, and thus that Kant's distinction between logical and physical modes of classification, as he conceived it, was no longer

<hr>

[75]Hettner, "Die Entwicklung der Geographie im 19. Jahrhundert," Geographische Zeitschrift, IV (1898), 305-320.

[76]Hettner, "Das Wesen und die Methoden der Geographie," ibid., XI (1905), 551n.

acceptable (p. 115).[77] Hettner was essentially an eclectic in philosophy; he adopted and adapted what he found useful in various philosophical positions. He gives us an important clue in his autobiography:[78]

It is good that I did not go into philosophy for many parts of that subject are alien to my intellectual individuality, and probably I would not have seen my way in modern philosophy. But the study of philosophy has been useful to me in so far as it gave my methodological endeavors a deeper foundation.

For instance, it appears that he worked out his own conception of the place of geography in relation to other sciences, and then sought philosophical grounding or justification for it in a combination of Auguste Comte and the neo-Kantians, Windelband and Rickert.[79] The remarks on Kant were added only in the form of welcome historical support.

With the approach of the centenary of Kant's death, there was for a time renewed interest in his geography. Gerland's article, published in Kant-Studien, was given originally as a course of twelve lectures in 1901. As mentioned earlier, much of this article was devoted to a critique of Kant's General Natural History and Theory of the Heavens, and of his early physical works. In fact, only one lecture was devoted to examination of the Physische Geographie, and much of it was concerned with a criticism of the obviously outdated content of the main body of the work. Only a few pages, consisting mainly of exposition without commentary, are devoted to the introduction to the Physische Geographie.[80] However, Gerland did approve strongly of Kant's ideas that "the history of the earth is nothing but a continuous geography," and "physical geography is the ground of history and of all other possible geographies"; and, of course, he criticized Kant's idea that "we have no natural history." Willy Kaminski's Ph.D. dissertation of 1905,[81] although in part

[77]The precise implication of this remark is far from clear, and Hettner does not enlarge on it. However, it appears to imply that the theory of evolution had achieved justification for a "logical" classification of biological phenomena, and hence that a "physical" classification could no longer be sharply differentiated from the former.

[78]"From My Life," see Schmitthenner, "Alfred Hettner," mimeo., p. 9.

[79]Hettner, Die Geographie, pp. 112-114.

[80]Gerland, "Immanuel Kant," 500-505.

[81]Über Immanuel Kants Schriften zur physischen Geographie. For a brief indication of the content of this thesis, see Hartshorne, "The Concept of Geography as a Science of Space, from Kant and Humboldt to Hettner," 104-105, 107.

concerned with examining Kant's influence and the claims of other referees to his geography, came to the conclusion that Kant's importance in geography was to be found in his teaching, particularly in his presentation of the nature of geography and its relation to the rest of knowledge. Kaminski appears to have been the first to determine that the core of Kant's ideas on geography was contained in the introduction to his Physische Geographie. [82]

The recent revival of interest in Kant's geography was mentioned in the introductory chapter. But it does appear evident that Kant's ideas have had more influence in the twentieth century than they ever did in the nineteenth, with the possible exception of their influence on Alexander von Humboldt. However, there are recent indications of a fairly widespread "positivist" revolt against Kant's concept of geography, or at least against the prevalent contemporary interpretation of that concept. Let us turn now to examine Kant's concept of geography in detail.

---

[82]Moreover, Kaminski may have been the first to have interpreted Kant as drawing a distinction between "systematic" or "subject matter" sciences, geographical or spatial sciences, and historical or temporal sciences ("Systematisch-begriffliche und geographisch-räumliche Anordnung. Die Klassifikation," Über Immanuel Kants Schriften zur physischen Geographie, chap. ix, pp. 67-71).

IV
Kant's concept of geography
(1) Geography as a science

A consideration of Kant's concept of geography can best proceed by way of an analysis of, and loose commentary on, the
introduction to his <u>Physische Geographie</u>. [1] Material from
other sources can be incorporated into the commentary as we
proceed. Topics that require discussion include geography as
a science, the relations between geography and anthropology
and between geography and history, geography as a propaedeutic and as an end-product of knowledge, geography and teleology, and the place of geography in a classification of the
sciences. But first, it will be necessary to have some idea of
the limits and scope of geography as conceived by Kant. Since
this topic is not discussed explicitly in the introduction to the
<u>Physische Geographie</u>, Kant's views must be reconstructed
from other sources.

THE LIMITS AND SCOPE
OF GEOGRAPHY

It is necessary to have a clear idea of the substantive limits
Kant set for the subject of geography, since commentators

---

[1]A translation of the introduction to the <u>Physische Geographie</u> is included as an appendix to this work.

have sometimes written on supposedly "geographical" works
by Kant that are in fact not even remotely geographical in content. Kant opens the main body of the Physische Geographie
with the remark that the surface (Oberfläche) of the earth is
to be divided initially into two main parts, the sea and continental land masses (p. 184). The notion of the "Oberfläche"
must be taken very seriously in Kant's work on geography, because he generally sticks quite close to this idea as setting the
bounds for geography. For instance, in introducing the topic
of earthquakes, Kant poses the question: "Is the cause of earthquakes to be sought on the surface of the earth, or deep in its
interior?" He answers the question by stating that: "In this
matter, physicists have not yet come to an understanding with
one another," but states his own opinion that "the cause is no
longer to be sought on the surface of the earth, but deep in
itself."[2] In discussing volcanoes, Kant indicates that in asking
about the origination of earthquakes, we must seek the opinions
of several physicists (p. 268). It is apparent, from these contexts, that the study of the causes of earthquakes and volcanoes,
which originate within the interior of the earth, is not the task
of the geographer but of the physicist. The geographer is concerned only with the effects or results, on the earth's surface,
of what happens in its interior, and he must accept the physicist's word on the latter. For instance, as Kant implies in
his discussion of volcanoes, the geographer is interested in the
fact that the stony soil on the slopes of volcanic Mount Vesuvius
can produce the grapes from which the fine Lacrima Christi
wine is made (p. 264). Thus, early works by Kant, which discuss earthquakes and volcanoes, are not to be construed as
works in geography, but as works in speculative physics of the
earth's interior.

Similarly, at the other end of the spatial spectrum, Kant's
work in speculative astronomy is not to be construed as work in
geography, despite the fact he indicates in the "Mathematische
Vorbegriffe" to his Physische Geographie that the geographer
requires a knowledge of astronomical principles insofar as he
is concerned with topics such as the seasons, the length of day
and night, the major zones of the earth's surface, longitude
and latitude, the effects of the sun and the moon on the earth,

[2]"Physische Geographie," GS, IX, 260-261.

etc. Astronomy and geography, in a remote sense, can be regarded as broadly similar, since astronomy, in one of its aspects, is concerned with the distribution of heavenly bodies in outer space, whereas geography is concerned with distribution on the surface of the earth. But Kant's work in speculative astronomy cannot be construed as work in geography, since the latter discipline is not concerned with outer space. Gerland[3] devoted well over half of his lengthy consideration of Kant's geography to a detailed discussion of the General Natural History and Theory of the Heavens, and to early works on the earth's interior. Such an undertaking could only have been based on a gross misunderstanding either of Kant's concept of geography or of the nature of geography itself.

However, the situation is much different respecting study of the atmosphere. In his first paper on that topic,[4] Kant points out that "one must think of the atmosphere as an ocean of elastic, liquid material, composed, so to speak, of layers of different density which decreases with greater height" (p. 491). Man lives at the bottom of this "ocean" of air, and is in the atmosphere and intimately related to it in a way that does not apply to the earth's interior or to outer space. Thus, the geographer must be concerned with the atmosphere in a way that he is not concerned with the earth's interior or with outer space. However, Kant presented this paper in the form of an announcement of his courses for the academic year 1756-57. It contains no mention of a course on physical geography, and the paper appears as an example of what would be discussed in "natural science." Kant makes it quite clear, though, that his theory is of basic geographical interest, for he indicates that once the information he was presenting had been acquired, the reader would derive much pleasure from looking at a map which showed the winds, both constant and periodic, of all climates, and the way in which they approached the coasts. In short, the reader would now be able to understand the motion of various winds over the surface of the earth (p. 501). In the following year, when he presented the first outline of his course of lectures on physical geography, he listed the atmosphere as an essential part of geographical study. In addition, he ap-

[3] "Immanuel Kant, " 1-43, 417-547.

[4] "Neue Anmerkungen zur Erläuterung der Theorie der Winde" (1756), GS, I, 489-503.

pended to that outline a brief essay on "Whether the West Winds in our Region are humid because they pass over a large Sea, "[5] thus indicating that study of the atmosphere and of climate is an intimate concern of geography. And so it remained through-out the history of his course of lectures on the subject. Kant's main work on the atmosphere, however, was primarily a con-tribution to speculative meteorology and not to climatology. In his day, the science of meteorology in its modern sense was just beginning to emerge, and he shows no indication of draw-ing a distinction between meteorology and climatology.[6]

Given Kant's concept of the limits of geography, it is ap-parent that his early physical works, aside from the papers on the atmosphere, contain little that can strictly be labelled geog-raphy. Kant's striking theory that frictional resistance between land masses and tidal currents on the earth's surface must cause a diminution of the earth's rotational speed, presented in his paper of 1754, "Whether the Earth has Undergone an Alteration in its Axial Rotation, "[7] since it deals with the earth-body as a whole, and its relations to other bodies in space, is a contribution to speculative astronomy and not to geography, although it is evidently of some geographical interest since the theory implies profound eventual changes on the earth's surface. Kant's other paper of 1754, "Die Frage, ob die Erde veralte, physikalisch erwogen, "[8] is of greater geographical interest be-cause Kant is concerned with the ageing or alteration of the earth from the standpoint of what evidence could be culled from observation of changes in its surface. Kant concludes in favour of the action of rain and streams:

In respect to changes of the earth's shape there remains to be discussed a single cause which can be counted upon with certainty; it consists in the

[5]Kant, GS, II, 10-12.

[6]Even Kant's main contribution to speculative meteorology, his general theory of the winds, was not as original as he had supposed ("Neue Anmerkungen zur Erläuterung der Theorie der Winde, " GS, I, 494). The topic had been somewhat inadequately dis-cussed by Halley and by Mariotte, with whose works Kant was acquainted, but he was apparently unaware of George Hadley's classic statement of 1735 as published in Vol. XXXIX of the Philosophical Transactions. For a brief history of the development of theory on the circulation of the winds, see Wolf, A History of Science, Technology and Philosophy in the 16th and 17th Centuries, I, pp. 316-320.

[7]Contained in Kant's Cosmogony, trans. by Hastie, pp. 3-15.

[8]GS, I, 193-213.

fact that the rain and streams continually attack the land and sluice it down from the highlands to the lowlands, gradually making the elevations into plains and, so far as in them lies, strive to rob the globe of its inequalities. This action is certain and no matter of opinion. The land is subject to this action so long as there is material on the declivities which can be attacked and transported by rain water.[9]

In this passage, Kant sides with the "gradualists," although he did not rule out the possibility of catastrophic changes altering the earth's surface. It is conceivable that collision with a comet, or a conflagration originating from within the earth's interior, could bring about a sudden change or even extinction of the earth (pp. 212-213).

Kant's General Natural History and Theory of the Heavens is, as he regarded it himself, a contribution to speculative astronomy and not to geography. As such, it contains virtually no geographical content. The heavenly bodies "form a co-ordinated whole, which is a World of worlds,"[10] and not just a "world," which is Kant's usual designation for the "whole" that is empirical geography. The only part of this work that can be construed as geography are some remarks in the introduction concerning land and sea breezes, and the general utility of various winds to mankind, which Kant introduced specifically as a common example of the proof of a benevolent Providence.[11] Kant nevertheless concludes his discussion with the remark:

[9]Trans. by S. G. Martin, "Kant as a Student of Natural Science," in Immanuel Kant: Papers Read at Northwestern University on the Bicentenary of Kant's Birth, pp. 101-111; quote on p. 105, from p. 209. Martin sees Kant, in this passage, "introduc[ing] in a very few words the conception of base-levelling, now familiar enough to geological students." However, there is nothing in the passage that could not have been written by Plato, Aristotle, or Posidonius, or even by Xenophanes. The idea of base-levelling did not receive adequate formulation until the work of William Morris Davis appeared at the beginning of this century ("Base-Level, Grade, and Peneplain," in his Geographical Essays, pp. 381-412).

[10]Kant's Cosmogony, trans. by Hastie, p. 64.

[11]Ibid., pp. 20-23. A few years later Kant repeated these arguments in much the same form in his own cosmological proof of the existence of God (GS, II, 97-98). The only other remarks in the General Natural History and Theory of the Heavens, so far as I can see, that bear even a remote relation to geography, conceived as the science of the earth's surface, are the following: brief mention of different kinds of matter of which the earth is composed (pp. 85-86); mention of Northern Lights and their possible benefit to inhabitants of the Arctic Zone (p. 102); the irregularity of the earth's surface, and the "fact" that these inequalities of the surface are produced more in the vicinity of the equator (pp. 110-111); some amusing and facetious remarks on what might have happened to the surface of the earth if that planet had had a ring such as

Of what utility are not the winds generally to the earth, and what use does not the acuteness of men make of them! Nevertheless no other arrangements are necessary to produce them than those general conditions of air and heat which must be found upon the earth even apart from these ends [p. 21].

Kant's three articles on the causes of earthquakes[12] are concerned primarily with the physics of the earth's interior. However, since the occasion for writing these articles was the disastrous Lisbon earthquake of 1755, Kant took the opportunity to make some remarks on the relations of man to natural catastrophes. Some of these remarks are of general geographical interest, since they throw light on Kant's concept of man in relation to the earth. In brief, the tenor of his remarks is as follows. Man must learn to adapt himself to nature and not demand that nature adapt to him. He can adapt by building better houses, and by aligning towns at right angles to the probable direction of the earthquake's cavity, but he is puny in the face of these great catastrophes. To regard these calamities as an expression of God's vengeance is criminal folly that arises from man's conceit in regarding himself as the sole purpose of nature. Man is only a part and not the whole of nature. In Nature's wider plan, earthquakes may be useful, and their usefulness to other aspects of nature may involve the sacrifice of human lives. In short, the earth cannot be preeminently regarded as the ideal dwelling-place for man (pp. 455-461).

Additional information on the question of the limits and scope of geography as conceived by Kant will be provided when we consider such topics as the relations between geography and physics, geography and anthropology, and geography and history, and more generally, the place of geography in a classification of the sciences.

Saturn's (pp. 129-131); mention of the fact that considerable portions of the earth's surface that man now inhabits are gradually being buried again in the sea while other regions are slowly reappearing (p. 149); and finally, some remarks on the "prodigality" of nature, on the abundance of its on-going creative power—innumerable multitudes of flowers and insects are destroyed in a single cold day, to be replaced in profusion by nature; and even man can look forward to his own extinction on earth (p. 150).

[12]"Von den Ursachen der Erderschütterungen"; "Geschichte und Naturbeschreibung der merkwürdigsten Vorfälle des Erdbebens"; "Fortgesetzte Betrachtung der seit einiger Zeit wahrgenommenen Erderschütterungen"; GS, I, 417-472.

GEOGRAPHY AS A SCIENCE

The consideration of Kant's concept of geography as a science may conveniently begin with some remarks on his distinction between rational, theoretical, and empirical sciences, [13] since this distinction appears, in most instances, to have been either ignored or misunderstood by geographers. To consider first Kant's distinction between rational and empirical, a priori or rational knowledge is knowledge that is absolutely independent of any experience. On the other hand, a posteriori or empirical knowledge is knowledge that is possible only through experience. [14] In a broad sense, Kant equates experience with empirical knowledge. [15] Experience, then, is knowledge that comes to us through the senses, under the forms of space and time. At an elementary level, it is knowledge of particular occurrences in space and in time. A priori knowledge, as knowledge that is independent of all experience, is supplied by human reason itself, and is characterized by universality and necessity. Such knowledge, then, in relation to science, is necessarily and universally valid, independent of the particularities of space and time. Empirical knowledge, on the other hand, since it lacks this element of universality and necessity, is always to some extent contingent and hypothetical, always dependent upon the limitations of actual experience.

The preceding remarks apply to knowledge. But how does Kant get from knowledge to science, and how does he distinguish between a priori sciences and empirical sciences? To consider the second question, that Kant intends such a basic distinction between the sciences is evident from his "Architectonic of Pure Reason," where, for instance, he proposes to remove empirical psychology from its traditional place within metaphysics, "the science which exhibits in systematic connection the whole body . . . of philosophical knowledge arising out of pure reason, "[16] and establish it eventually "in a complete

[13] Kant's basic distinction is between rational and empirical sciences; he rarely uses the concept "theoretical science." Nevertheless, it is apparent that he intends a three-fold distinction among what he calls the sciences. What Kant occasionally refers to as "empirical" physics, I have designated as theoretical in my exposition of his classification of the sciences, which will be outlined in chapter vi.

[14] Kant, Critique of Pure Reason, B3 ff.

[15] Ibid., B147, B166.

[16] Ibid., A841 = B869.

anthropology, the pendant to the empirical doctrine of nature."[17]
Kant here sets the "empirical doctrine of nature," and not
geography, alongside anthropology, although the introduction
to the Physische Geographie clearly indicates that geography
and anthropology together form a "whole" of our empirical
knowledge of the world. Strictly speaking, since geography
confines itself to the surface of the earth, presumably the
"empirical doctrine of nature" is broader than geography, because it would also embrace whatever direct empirical knowledge we had of the heavens and of the earth's interior.
However, for our purposes, this difference is a minor one.

In the broadest sense, Kant divides the sciences into empirical, based on knowledge from data, and rational, based on
knowledge from a priori principles.[18] Empirical science, as
we have noted, is divided into geography and anthropology, rational science into mathematics, based on the construction of
concepts, and philosophy, or metaphysics in its broadest sense,
which in turn is divided most generally into rational physics
and rational psychology.

Kant also contrasts science with "common knowledge":

> Science is contrasted with common knowledge, that is, the content of a
> cognition, as a mere Aggregate. System rests on an idea of the whole,
> which precedes the parts; whereas, in common knowledge, or the mere
> aggregate of cognitions, the parts precede the whole. There are Histori-
> cal sciences and Rational sciences.[19]

Common knowledge, since it is concerned only with the content
of cognitions, or with cognitions in their isolation from one
another, can never be more than a mere aggregate of information. On the other hand, empirical science qualifies as science by virtue of its systematic character. It is essentially
the idea of system that elevates knowledge to the status of science. This introduces an a priori element into empirical science, which will be discussed later in this chapter. In the
strict sense, for Kant, no science can be merely empirical;
it has to contain some a priori elements or ideas. This situation serves also to complicate the neat distinction drawn above

[17]Ibid., A849 = B877.
[18]Ibid., A836 = B864.
[19]Introduction to Logic, p. 62. Throughout the Introduction to Logic, Kant employs
the term "historical" in its traditional meaning of empirical, or knowledge derived
from data gained through sense experience.

between mutually exclusive empirical and a priori sciences.

That Kant did not intend an absolutely rigid distinction between rational or philosophical and empirical sciences is perhaps shown in the following remark: "All philosophy is either knowledge arising out of pure reason, or knowledge obtained by reason from empirical principles. The former is termed pure, the latter empirical philosophy."[20] If we consider philosophy "in the cosmic conception of it," rather than "in the scholastic conception of it" as we do in the "Architectonic of Pure Reason," it then becomes the science of "the use of our reason," "of the relation of all knowledge and every use of reason to the ultimate end of human reason." In this sense, the ultimate philosophical question is "What is Man?" and the answer to this question is sought in anthropology.[21] Anthropology is an empirical and pragmatic discipline, and hence the empirical and the pragmatic is a vital element in Kant's philosophy. Although Kant talks of a "physiological" anthropology and contrasts it with a "pragmatic" anthropology,[22] he presents us with only a pragmatic anthropology. The term "physiological" in this context, meaning "what nature makes of man," as contrasted with "pragmatic," meaning "what he as free agent makes of himself," does not necessarily imply a theoretical knowledge of man. In the next chapter some evidence will be produced to support the contention that "physiological" anthropology is in part the task of geography. As such, it remains an empirical undertaking.

However, with regard to the question of the distinction between theoretical and empirical sciences, the Critique of Pure Reason aims, in part, at answering the questions: "How is pure mathematics possible?" and "How is pure science of nature possible?" These sciences actually existed in Kant's day, in the form of Euclidean geometry and Newtonian physics.[23] Thus, part of the task of the Critique of Pure Reason is to

[20]Critique of Pure Reason, A840 = B868.

[21]Kant, Introduction to Logic, p. 15; see also Kant's letter of May 4, 1793, to Karl Stäudlin, Kant: Philosophical Correspondence, pp. 205-206. However, for Kant, all science and all philosophy contributes to the ultimate solution of the question, "What is Man?"

[22]"Anthropologie," GS, VII, 119.

[23]Critique of Pure Reason, B20.

provide the philosophical justification or rational grounding for the established science of the day. However, Kant speaks occasionally of "empirical physics."[24] Physics is, in part, an empirical science, since the discovery of actual scientific laws depends upon experience. Kant never tires of pointing out that we cannot simply deduce the actual laws of nature from a priori principles.[25] In addition, those principles, as employed in scientific contexts, can apply only to possible experience, and moreover, in the final analysis, to possible experience in the sense of its actualization. For instance, Kant cites the case of the possibility of inhabitants on the moon. Such a possibility must be admitted, but to date, since we have no experience of such beings, the possibility of experience can mean only that we may encounter them in the course of the empirical advance of our knowledge.[26] On the other hand, we may never encounter such beings in actual experience. Physics, however, as empirical and as a fully developed science, rests upon a set of a priori principles, such as those pertaining to the permanence of matter, to inertia, to the equality of action and reaction, to causality, etc.[27] Although we may separate the a priori and empirical parts of natural science, and treat them separately—the one as metaphysics, the other as particular empirical science—natural science proper, i.e., physics, always presupposes metaphysics.[28] In short, the two aspects are united in a fully developed theoretical physics or natural science.

Not all natural sciences, of course, have attained the strict scientific status of physics. As Kant points out:

Everything in nature, whether in the animate or inanimate world, takes place according to rules, although we do not always know these rules. . . . All nature, indeed, is nothing but a combination of phenomena which follows rules; and nowhere is there any irregularity. When we think we find such, we can only say that the rules are unknown.[29]

---

[24]For instance, see Introduction to Logic, p. 57; and Prolegomena to Any Future Metaphysics, sec. 15.

[25]For instance, see Critique of Pure Reason, B165.

[26]Ibid., A493 = B521.

[27]Ibid., B21; Prolegomena to Any Future Metaphysics, sec. 15.

[28]Kant, Metaphysical Foundations of Natural Science, pp. 139-140.

[29]Introduction to Logic, p. 1.

Although we do not know all the rules, principles, and laws of nature, we approach nature, in strictly scientific studies, from the point of view that such laws exist and can be discovered. For instance, the chemistry of Kant's day consisted only of empirical facts and laws. Since it lacked an adequate theoretical grounding in a priori principles, it should, strictly speaking, be termed "systematic art" rather than science.[30] Chemistry was not then a theoretical science, but part of the "empirical doctrine of nature." But, if we are to distinguish chemistry from geography, the true "empirical doctrine of nature," then presumably chemistry is capable of becoming a theoretical science, i.e., of being grounded on a set of a priori propositions. In order to maintain the distinction between theoretical and empirical sciences, or even sciences that are capable of becoming theoretical, such as chemistry, then geography, by its very nature, must be an empirical science, a science incapable in principle of becoming theoretical, or of being grounded on a set of thoroughgoing a priori propositions. Kant points out in the introduction to his Physische Geographie that geography was not a well-developed science in his day, and so presumably our geographical knowledge of the world is capable of becoming far more extensive. Nevertheless, it would remain an essentially empirical science.

If the distinction between theoretical and empirical sciences is to be maintained as basic, then concepts such as "nature" and "world," as employed in geography, must have a different meaning than they have in physics, and in a priori grounded sciences generally. In one sense of the term, "Nature is the existence of things, so far as it is determined according to universal laws."[31] Since "universal laws" can only be supplied a priori by reason, this is the rational concept of nature. However, at the same time, this concept of nature is a thoroughly mathematical one:

A pure philosophy of nature in general, namely, one that only investigates what constitutes a nature in general, may thus be possible without mathematics; but a pure doctrine of nature respecting determinate natural things (corporeal doctrine and mental doctrine), is only possible by means of mathematics; and as in every natural doctrine only so much science proper is to

---

[30]Kant, Metaphysical Foundations of Natural Science, p. 138.
[31]Kant, Prolegomena to Any Future Metaphysics, sec. 14.

be met with therein as there is a cognition a priori, a doctrine of nature can only contain so much science proper as there is in it of applied mathematics.[32]

Although we may think of nature in a purely rational sense, insofar as a study is to qualify as "pure" science of nature its propositions must be expressed in mathematical form. The thoroughgoing application of mathematics introduces the theoretical aspect of the study of nature.

Kant, however, points out that

the word "nature" assumes yet another meaning, which determines the object, whereas in the former sense it only denotes the conformity to law of the determinations of the existence of things generally. If we consider it materialiter (i.e., in the matter that forms its objects) "nature is the complex of all the objects of experience."[33]

In this sense, we are concerned with nature as an actuality which can be confirmed by experience, yet, at the same time, with it as it necessarily conforms to a priori laws. This concept of nature introduces the empirical aspect of physics. We cannot simply deduce the actual laws of physics from a priori principles; we must discover them in nature. Yet, these laws, as "universal" laws of "empirical" physics, must be exhibited in their conformity to universal and necessary a priori laws of reason. However, this concept of nature, "the complex of all the objects of experience," is essentially the same concept of nature as employed in geography, where it is expressed as "the world [surface of the earth] as the sum-total of all knowledge of experience."[34] How, then, do we distinguish geography as the "empirical doctrine of nature" from "empirical" physics? This is not an easy question to answer, since, in the body of the Physische Geographie, Kant often discusses the causes (Ursachen) of various phenomena occurring on the sur-

---

[32]Kant, Metaphysical Foundations of Natural Science, pp. 140-141.

[33]Prolegomena to Any Future Metaphysics, sec. 16; see also Critique of Pure Reason, A216 = B263, for a statement by Kant respecting nature in the "empirical sense."

[34]Nevertheless, there is evidently a difference between "objects of experience" and "knowledge of experience." Conceived as a purely academic discipline, geography is concerned with "knowledge" rather than with "objects." But conceived as a research discipline, geography, as well as "empirical" physics, is concerned with "objects of experience." In this context, I am concerned with geography as a research discipline; its role as an academic discipline will be discussed later when we come to consider geography as a propaedeutic and as an end-product of knowledge.

face of the earth.[35] Evidently then, geography as a "description of nature" is not a description in the "mere description" sense of that term. It is to some extent a law-finding, and not exclusively fact-finding, science. Moreover, even to discuss causality is to introduce one of the categories of the understanding, and hence to introduce a rational or a priori element into geography, for "categories are concepts which prescribe laws a priori to appearances, and therefore to nature, the sum of all appearances."[36]

Kant nowhere appears to give a satisfactory answer to this question, although he obviously intends that geography studies nature and the world differently than does physics.[37] He does, however, on at least one occasion discuss "special laws":

> Pure understanding is not, however, in a position, through mere categories to prescribe to appearances any a priori laws other than those which are involved in a nature in general, that is, in the conformity to law of all appearances in space and time. Special laws, as concerning those appearances which are empirically determined, cannot in their specific character be derived from the categories, although they are one and all subject to them. To obtain any knowledge whatsoever of these special laws, we must resort to experience.[38]

A distinction can perhaps be drawn between types of "special laws." For instance, in discussing aspects of "empirical" physics of which "we have not, a priori, the least conception," Kant couches his discussion in terms of general concepts such as "forces," "motions," "alterations," "bodies," and

---

[35]For instance, he discusses the causes of the saltiness of water (pp. 198-199); of different kinds of salt (p. 201); of waves (pp. 208-210); of breakers (p. 211); of ocean currents (pp. 212-213); of currents in the Mediterranean Sea and in the Gulf Stream (p. 215); of whirlpools (p. 215); of tides and their ebb and flow (p. 219); of sand-banks (pp. 239-240); and of earthquakes (pp. 260-263).

[36]Kant, Critique of Pure Reason, B163. See also B234, where Kant argues that "Experience itself—in other words, empirical knowledge of appearances—is thus possible only in so far as we subject the succession of appearances .... to the law of causality; and, as likewise follows, the appearances, as objects of experience, are themselves possible only in conformity with the law." This statement appears to indicate that only law-finding science can qualify as science for Kant; a merely fact-finding "science" could not so qualify.

[37]Perhaps Kant's clearest statement respecting this intention occurs in his letter of late 1773 to Marcus Herz (Kant: Philosophical Correspondence, p. 79), referred to previously, where he remarks that "anthropology . . . along with physical geography and distinct from all other learning, can be called knowledge of the world."

[38]Critique of Pure Reason, B165.

"states. "[39] On the other hand, in law-finding contexts in geography, the discussion proceeds in terms of particulars such as salt, seas, waves, breakers, ocean currents, sand-banks, etc. "Empirical" physics, then, deals with "universal" and abstract categories of phenomena, geography with specific and concrete categories. This suggests a further distinction; although it is not one that Kant makes, it is, I think, implicit in his presentation. The laws of geography are "regional, " in Mehlberg's sense of the term;[40] the laws of "empirical" physics are "universal, " insofar, of course, as it is possible to attain to universal laws on empirical grounds alone. In dealing with waves, breakers, ocean currents, sand-banks, etc., the geographer is dealing with phenomena that are not uniformly or universally distributed over the surface of the earth, but are "regional, " confined to definite areas of space and destined to persist over only limited periods of time. Thus, the laws of geography, being more specific and concrete, are contingent in a sense in which the laws of "empirical" physics are not.

Similarly, Kant uses the concept "world" in a variety of senses. In strictly scientific contexts, it signifies "the mathematical sum-total of all appearances. "[41] In geographical contexts, as we have seen, the term "world" means the "sumtotal of all knowledge of experience" (of the earth's surface). The "world-whole, " as Kant makes quite clear, is only an "Idea" (A408 = B434), a concept of reason that we impose a priori upon nature. We can never arrive at this Idea on the basis of experience alone.

If I represent the earth as it appears to my senses, as a flat surface, with a circular horizon, I cannot know how far it extends. But experience teaches me that wherever I may go, I always see a space around me in which I could proceed further; and thus I know the limits of my actual knowledge of the earth at any given time, but not the limits of all possible geography [A759 = B787].

But,

even if we suppose the whole of nature to be spread out before us, and that of all that is presented to our intuition nothing is concealed from our senses

[39]Ibid., A207 = B252.

[40]The Reach of Science, pp. 60, 161 ff. Regional laws "hold only within a region confined to the surface of the earth during a finite time-interval" (p. 60).

[41]Critique of Pure Reason, A418 = B446.

and consciousness, yet still through no experience could the object of our ideas be known to us in concreto [A482 = B510].

In a mathematical sense, of course, we know the limits of the earth, although we are inherently ignorant of its extent as the "sum-total of all knowledge of experience":

But if I have got so far as to know that the earth is a sphere and that its surface is spherical, I am able even from a small part of it, for instance, from the magnitude of a degree, to know determinately, in accordance with principles a priori, the diameter, and through it the total superficial area of the earth; and although I am ignorant of the objects which this surface may contain, I yet have knowledge in respect of its circuit, magnitude, and limits [A759 = B787].

The Idea of a "world-whole" is used in several other contexts by Kant. It is, for instance, as a consequence or an extension of the principle of community, [42] a postulate of empirical thought. [43] Kant's discussion here refers to "empirical" physics, but it is evidently applicable, in part, to geography, insofar as that discipline is to be regarded as supplying systematic knowledge. Since geography is concerned with studying the relations among things that coexist beside one another in the space of the earth's surface, it obviously presupposes the principle of community. The "world-whole" or the "world in general" is, however, only a regulative Idea (A684 = B712). It is regulative in the sense that we regard the world as forming a systematic whole, although the Idea can never be derived from experience. The Idea is not constitutive of nature since we impose it on nature, approach nature from a certain perspective, rather than finding the systematic unity in nature itself.

At this point we may conveniently introduce the ideas "architectonic" and "system," since they are mentioned by Kant in the introduction to the Physische Geographie. Both concepts are applicable to geography as an empirical science.

[42]Ibid., A211. "All substances, so far as they coexist, stand in thoroughgoing community, that is, in mutual interaction."

[43]Ibid., A218n = B265n. "The unity of the world-whole, in which all appearances have to be connected, is evidently a mere consequence of the tacitly assumed principle of the community of all substances which are coexistent. For if they were isolated, they would not as parts constitute a whole. . . . We have, however, . . . shown that community is really the ground of the possibility of an empirical knowledge of coexistence."

By an architectonic I understand the art of constructing systems. As systematic unity is what first raises ordinary knowledge to the rank of science, that is, makes a system out of a mere aggregate of knowledge, architectonic is the doctrine of the scientific in our knowledge.

Kant goes on to define "system" as follows:

By a system I understand the unity of the manifold modes of knowledge under one idea. This idea is the concept provided by reason—of the form of a whole—in so far as the concept determines a priori not only the scope of its manifold content, but also the positions which the parts occupy relatively to one another. [44]

This introduces the idea of a plan or arrangement of knowledge. Kant's illustrations of this plan of reason are fairly straightforward. For instance, the example he gives in the introduction to the Physische Geographie, that of building a house, points out that one must first have an idea or plan of the whole house, and of the arrangement of the various rooms, etc., before he commences to build the house. A more geographical illustration is the following:

Towns, for example, tend to become . . . aggregates, if, without the supervision of the police [or authorities], a plot of land is divided among a number of builders, each working for himself according to his own opinions. If, however, the idea of a whole according to a certain principle can and ought to be presupposed before the determination of the parts, the division must then take place in a scientific manner. [45]

Obviously, we cannot build houses without plans, nor can we have town-planning unless we first construct a prearranged plan for a town.

But how, specifically, does the foregoing apply to geography as a science? Initially, the examples imply that there must be a fair degree of consensus respecting the limits, scope, and content of a discipline amongst its practitioners if that discipline is to be recognized as scientific. In addition, geography possesses the first requirement for a system, the idea of a whole—in the case of geography, the world understood as the surface of the earth. This whole sets the limits within which geography is to be conceived. Geography also

[44]Critique of Pure Reason, A832 = B860.
[45]Kant, On Philosophy in General [First Introduction to the Critique of Judgment], trans. by Humayun Kabir, p. 81.

contains systematic elements in that it can show connections between wind and waves, waves and sand-banks, etc. It is not, however, a systematic science in the sense in which physics is. The latter science can exhibit its propositions in thorough-going systematic interconnection, which geography cannot do. Geography cannot, for instance, show any essential connection between many diverse parts or features of the earth's surface. In contrasting the "description of nature" with, on the one hand physics as a theoretical science, and on the other hand the "history of nature," Kant indicates that the "description of nature," as distinct from the "history of nature," approaches the status of physics since it "presents itself as a science in all the splendour (<u>Pracht</u>) of a great system."[46] Evidently, geography has some of the appearances of a system, but hardly the thorough-going systematic character of physics. What precisely Kant means, however, when he claims we "derive" the parts from the "whole," or that the "whole" determines the "positions which the parts occupy relatively to one another," is not altogether clear. What he seems to have in mind is that we organize the "whole," and thus inevitably impose upon our experiences some organizational structure or schema, which entails logical restrictions as to the manner in which the "parts" are to be related to one another; we are dealing here evidently with a whole inclusive of parts relationship, and not with a whole independent of parts relationship. In addition, the structure is not derivable from experience. Kant does point out, though, that in the development of a science the schema, and even the idea of the discipline which its founder originally had, usually undergoes considerable change.[47] Hence, as regulative of ideas, and not constitutive of nature, there is nothing inherently necessary about any particular schema adopted for organizing the material of a specific empirical science. Similarly, in planning a house or a town, any number of specific plans are possible.

At an elementary or concrete level, the geographer classifies or groups phenomena under concepts such as land, sea,

---

[46]"Über den Gebrauch teleologischer Principien in der Philosophie," <u>GS</u>, VIII, 162. However, there appears to be an element of sarcasm in Kant's use of the word "Pracht" in this context.

[47]<u>Critique of Pure Reason</u>, A834 = B862.

mountains, rivers, animals, plants, etc. At a more abstract level, he organizes his material in terms of concepts such as physics, mathematics, morals, politics, commerce, theology, and region. However, geographical material is also planned or arranged in a much broader sense. As Kant remarks in the introduction to his Physische Geographie:

Physical geography is thus the first part of knowledge of the world. It belongs to an idea (Idee) which is called the propaedeutic to understanding our knowledge of the world. Instruction in it still appears to be very defective. Nevertheless, it is this knowledge that is useful in all possible circumstances of life. Accordingly, it is necessary to acquaint oneself with it as a form of knowledge that may subsequently be completed and corrected by experience.

We anticipate the future experience, which we afterwards have of the world, through instruction and a general summary of this kind which gives us, as it were, a preliminary idea (Vorbegriff). We say of the person who has travelled much that he has seen the world. But more is needed for knowledge of the world than just seeing it. He who wants to profit from his journey must have a plan beforehand, and must not merely regard the world as an object of the outer senses.

Geography provides us with a prearranged plan for knowledge of the world, through which, to some extent, we can anticipate our future experience of the world. But this knowledge is always corrected, or made more exact, through actual experience. Geographical knowledge, then, in this cosmic concept of it, is "rough." This introduces another important respect in which geography differs from "empirical" physics. Kant distinguishes between "rough" and "exact" knowledge. [48] "Rough" knowledge is characteristic of "historical"—i.e., empirical—sciences, and contains a margin of error. In this class of sciences, a wider determination for cognitions is permissible, and it may "include errors which do not interfere with its purpose." In the case under consideration—geography in its cosmical concept—this latitude is permissible since the purpose of geography is to provide a meaningful arrangement or "preliminary idea" of all our knowledge of the world. Because of its cosmical empirical extent, geography is characterized by what Kant calls the "faith of history" or the "so-called historical belief." [49] Since the experience of any

[48]Introduction to Logic, pp. 45-46.

[49]Kant, "Critique of Teleological Judgement," Critique of Judgement, trans. by James C. Meredith, Part II, p. 143; see also Introduction to Logic, pp. 58-63.

given individual is very limited, in order to arrive at a knowledge of the world as the "sum of all appearances," the geographer must rely upon the testimony of others. This testimony,
however, must be reliable; in this sense, the "faith of history"
is no matter of faith at all, but is "matter of fact," since the
reliable facts of history and geography were the "personal
experience and matter of fact" of some particular witness.
Presumably one of the tests of reliability is that the witness be
able to present his observations in a systematic and organized
fashion, guided by principle and method. On these grounds,
Kant was able to declare that he had "no use for the merely
empirical traveller and his tales."[50] On the other hand, the
physicist, in arriving at his view of the world as the "sum of
all appearances," since he is concerned with highly abstract
matters, need not necessarily rely on the testimony of others.

With the foregoing information in mind, let us turn to the
introduction to the Physische Geographie, and Kant's contrast
between the "system of nature" and geography.[51] Basically,
however, such a contrast is misleading; since geography deals
with nature and is systematic to some extent, it can also be
regarded as a system of nature. Similarly, physics is also a
system of nature. What Kant is contrasting essentially are
different systems of nature. The distinction between Linnaeus'
"system of nature" and geography is made in terms of a con-

---

[50]"Über den Gebrauch teleologischer Principien in der Philosophie," GS, VIII, 161.

[51]This contrast has often been badly misconstrued by geographers, who have interpreted Kant as drawing a distinction between physics, or "systematic" sciences generally, and geography (see for instance, Hartshorne, The Nature of Geography, pp.
134-135, 140, 146; and "The Concept of Geography as a Science of Space, from Kant
and Humboldt to Hettner," 99). Too much has been made of Kant's statement in the
introduction to Physische Geographie that, "The classification of knowledge by concepts is the logical, that by time and space the physical classification." Taken literally, on the basis of Kantian philosophy, this statement is meaningless, since time
and space are forms of intuition not things, and since obviously, geography, just as
any other science, requires concepts. One can only suppose that Kant is writing very
loosely in this context for introductory purposes. In a famous passage in the Critique
of Pure Reason (A51 = B75), Kant declares that, "Thoughts without content are empty,
intuitions without concepts are blind." Later in the same work (A258 = B314), Kant
remarks that, "Understanding and sensibility, with us, can determine objects only
when they are employed in conjunction. When we separate them, we have intuitions
without concepts, or concepts without intuitions—in both cases, representations which
we are not in a position to apply to any determinate object." "Empirical" physics, as
well as geography, requires space and time as much as it requires concepts. Hence,
by the notion of a purely "logical" classification of knowledge, Kant cannot have physics
in mind.

trast between logical division and physical division of a "whole."
Kant mentions this distinction in the Critique of Pure Reason,[52]
although he sheds little further light on it. A logical division
of nature, through organizing phenomena in terms of genera,
species, subspecies, etc., aims at the unity of a system.
Thus, Linnaeus sought to organize the whole of nature into an
hierarchical system under the categories: empire, kingdom,
class, order, genus, species, and subspecies or variety.
These concepts are logical, that is, they bear no direct relation to observable entities, or, as Kant puts it, "this is a
classification that I make in my head."[53] On the other hand,
a physical division of nature, through concentration on observable entities, or on entities derivable from observation,
seeks to secure the widest possible empirical coverage and
spatial differentiation within the system. Thus, the surface
of the earth, regarded as a physical "whole," could be sub-
divided into a number of parts—mountains, rivers, lakes,
etc.—which presumably could be further subdivided into

smaller and smaller "parts." However, the distinction as
drawn requires several qualifications. One could, for instance,
operate within Linnaeus' system, or one like it, and yet concentrate one's attention upon the diversity of species, and upon
subspecies or varieties. Evidently, then, concentration upon
either the unity or the diversity of the system of nature is possible within its logical conception. These approaches stem
from different "interests of reason":

This twofold interest manifests itself also among students of nature in the
diversity of their ways of thinking. Those who are more especially speculative are, we may almost say, hostile to heterogeneity, and are always
on the watch for the unity of the genus; those, on the other hand, who are
more especially empirical, are constantly endeavouring to differentiate nature in such manifold fashion as almost to extinguish the hope of ever being
able to determine its appearances in accordance with universal principles.[54]

These "maxims of reason" both involve legitimate approaches
to the study of nature. Hence, to argue for the retention of
one approach to the total exclusion of the other would be to

[52]A655 = B683.
[53]"Physische Geographie," GS, IX, 160.
[54]Critique of Pure Reason, A655 = B683.

commit a serious "category mistake"[55]—the mistake of taking as constitutive of nature what is really only regulative of ideas, and hence without objective grounding.[56]

The physical division of nature, since it embodies a more empirical approach, would thus concentrate primarily upon the diversity and heterogeneity of nature. One could presumably subdivide the surface of the earth, the "whole" with which one starts, into finer and finer entities. But in neither the logical nor the physical division of nature will one ever arrive at the totally individual or unique; further subspecies or varieties are always assumed to be contained under any species.[57] In this sense, no science, theoretical or empirical, is ever concerned with the unique.

> Could Linnaeus have hoped to construct a system of nature if he had had to ensure that if he found a stone which he called granite this would be distinguished by its inner nature from every other apparently identical one? He would thus invariably expect to encounter only individual things isolated for the understanding, but never a class of them which could be brought under genus-species concepts.[58]

Similarly, geography as a system of nature is not concerned with the unique. The unique is the concern of "common knowledge." Although theoretically the geographer could subdivide the surface of the earth into ever more minute "parts," in practice the extent of the subdivision is governed by the cosmical concept of geography. In its cosmical concept, geography aims at utility, at "that in which everyone necessarily has an interest."[59] This is a vague criterion, but at least it guarantees that the geographer does not descend into trivia in his physical division of nature. For instance, preoccupation with someone's front lawn would be geographically trivial, whereas concern with the major grasslands of the world, and their potential to serve human needs, would be geographically significant.

---

[55]I am borrowing Gilbert Ryle's expression (The Concept of Mind, pp. 16 ff.).

[56]Critique of Pure Reason, A666-A668 = B694-B696.

[57]Ibid.

[58]Kant, First Introduction to the Critique of Judgment, trans. by James Haden, p. 20n.

[59]Kant, Critique of Pure Reason, A840 = B868; Introduction to Logic, p. 14. Moreover, geography is essentially a pragmatic discipline ("Von den verschiedenen Racen der Menschen," GS, II, 443n).

There are, of course, essential differences between a logical division and a physical division of nature. To employ one of Kant's examples—drawn from his introduction to Phys-ische Geographie—in Linnaeus' "system of nature," crocodiles and lizards are grouped together, since they are members of the same class of animals. The systematic character of the classification is revealed in the uniting of the two animals under the same class. On the other hand, in the physical division of nature, which is concerned with specific conditions of space and time, crocodiles and lizards are treated separately, since they are found in different places on the earth's surface. Nevertheless, the geographer is concerned with crocodiles and lizards as species, and not with this or that particular crocodile or this or that particular lizard. Similarly, he is concerned with mountains, rivers, lakes, etc., generally, and not merely with specific examples of these classes of phenom-ena. Although Kant is never very clear about it, presumably the systematic element in the physical division of nature would be indicated by showing that, for instance, crocodiles and liz-ards are associated with and dependent upon fairly specific physical environmental conditions which were different for each species of animal.

Kant, as we have already seen, rejected Linnaeus' "sys-tem of nature" as it stood, although he would have regarded as legitimate the attempt to create a "logical" system of nature.[60] Kant rejected it on a variety of grounds—it lacked the idea of a "whole," it had no adequate theoretical grounding, its empir-ical warranty was questionable, and it lacked any ultimate aim or utility beyond that of merely classifying for the sake of classifying. In short, it lacked explanatory power, and hence remained a scholastic "description of nature." On the other hand, although its basic elements are abstract concepts and not concrete entities or purely logical constructs, physics, as a "system of nature," possesses explanatory power—and hence empirical warranty—since it can predict under specific given conditions of space and time.

The geographer, as noted earlier, also organizes material in terms of more abstract or logical concepts such as physics,

[60]This is in fact attempted, I think, in the First Introduction to the Critique of Judg-ment. We shall have something to say on this matter when we discuss "geography and teleology" in a later chapter.

mathematics, morals, politics, commerce, theology, and region. The mathematical view of the "world-whole," of course, rests on a priori grounds, and is thus a theoretical, not an empirical, science. With respect to the other concepts, they do not appear to form an hierarchical order, except for the fact that physical geography is the basis of all other possible geographies. In this sense, the other geographies bear some systematic relation to physical geography, although apparently not to each other, except in the case of regional geography, wherein the interconnections among various types of phenomena are studied in specific regional or national contexts. Kant says virtually nothing about regional geography in the introduction to the Physische Geographie, beyond mentioning it as one of the approaches employed in the subject, although, as we have seen, he certainly suggested this role for regional geography in the earliest announcement of his course, in 1757. At no time, however, does Kant appear to have devoted much space to moral, political, commercial, and theological geography. Although abstracts of these various geographies continued to be presented in the introduction to the Physische Geographie, he apparently never found the time to organize and present material on them. His geography remained essentially and primarily a physical geography. Since physical geography and physics, by and large, both study the same objects (with the provision that geography is confined to the study of physical phenomena occurring on the surface of the earth),[61] the former could be described, and was in fact presented by Kant, as a concrete and largely empirical system of nature, whereas the latter could be described as an abstract and theoretical system of nature.

[61]See especially the quotation used to introduce the next chapter, where Kant specifically indicates that the objects studied by physics and geography are much the same.

V

Kant's concept of geography (2) Geography in relation to the empirical sciences, anthropology and history

GEOGRAPHY AND ANTHROPOLOGY

The physical geography, which I herewith announce, belongs to an idea (Idee) which I create for myself for purposes of useful academic instruction, and which I could call the preliminary exercise in the knowledge of the world. This knowledge of the world is that which serves to procure the pragmatic for all the acquired sciences and aptitudes. It will be useful not only for school but also for life, and through it the apprentice who has finished his training can be introduced on to the stage of his destination, which is the world. Here before him lies a twofold field, namely nature and man, of which he has a plan for the time being through which he can put into order, according to rules, all his future experiences. Both parts, however, have to be considered cosmologically, that is, not according to what their objects contain as peculiar to themselves (physics and empirical knowledge of soul), but what their relationship is in the whole in which they stand and in which each has its own position. The first form of instruction I call physical geography . . . the second anthropology. [1]

---

[1]Kant, "Von den verschiedenen Racen der Menschen," GS, II, 443n.

This passage is perhaps the clearest statement by Kant on the close relation between geography and anthropology. Together they form a "whole" of our pragmatic, empirical knowledge of the world. Both are pragmatic, in that each provides knowledge that is useful for life. Anthropology, however, is pragmatic in the further sense that it purports to give insight into "what man, as free agent, makes of himself. " Geography deals with nature, anthropology with man or soul. But there is evidently an overlap here, since geography also deals with man, insofar as he is to be regarded as a part and product of nature. Anthropology, then, deals essentially with man conceived as soul or self, with what, in some sense, is essential to man's "inner" constitution, but only, of course, in a pragmatic and empirical way. As Kant points out, the question whether or not man has a soul, conceived as an immaterial substance, does not arise in anthropology. This is a theoretical question, not a pragmatic and empirical one. Nevertheless, man does believe, psychologically, that he possesses a soul, and that the soul is the organ or object of inner sense.[2] The soul is an immaterial substance that "resides in a place of a smallness impossible to describe."[3] Hence, since the soul appears to reside at a point, "one cannot assign a relation in space to what is determinable only in time."[4] Consequently, since the soul is a purely temporal phenomenon, knowledge of its appearances can be had only through inner sense.

This raises the issue of Kant's distinction between outer and inner sense. This distinction is of crucial importance for his separation of anthropology from geography, since the world as the object of outer sense is nature, and hence the concern of geography, whereas the world as the object of inner sense is man conceived as soul or self, and is the concern of anthropology. Yet, Kant's doctrine of outer and inner sense remains extremely obscure. The doctrine of outer sense, however, presents no particular difficulty. It refers to the perception or intuition of things external to ourselves, in space, as it comes

---

[2]Kant, "Anthropologie, " GS, VII, 161.

[3]Quoted in Sir Charles Sherrington, Man on His Nature, p. 253; quoted from the Hartenstein edition of Kant's works (1867), II, p. 332.

[4]Quoted in Sherrington, Man on His Nature, p. 202; quoted from the Hartenstein edition of Kant's works (1839), X, p. 112.

to us through the five "outer" senses—sight, hearing, smell, touch, and taste—of which the most important, for purposes of knowledge of nature, is sight. To each individual, the activity or behaviour of all other persons is revealed to him through the outer senses, and in this respect, human behaviour is to be understood as entirely natural. In addition, Kant is quite clear about the fact that knowledge of my own body is revealed to me through the outer senses.[5] Hence, knowledge of the human body is also entirely natural. On these grounds, the study of race, since it is based upon perception of bodily characteristics, is the concern of geography and not of anthropology.

It is the doctrine of inner sense that presents difficulty. I am far from confident that I have more than an elementary grasp of what Kant is trying to say, but an attempt must be made to make some sense of the doctrine.[6] At the empirical level (and, since geography and anthropology are empirical disciplines, this is essentially the level which concerns us here), Kant evidently regards the distinction between outer and inner sense as one that we are necessarily aware of in our own immediate experience:

If then we ask, whether it follows that in the doctrine of soul dualism alone is tenable, we must answer: "Yes, certainly; but dualism only in the empirical sense." That is to say, in the connection of experience matter, as substance in the [field of] appearance, is really given to outer sense, just as the thinking "I," also as substance in the [field of] appearance, is given to inner sense. Further, appearances in both fields must be connected with each other according to rules which this category introduces into that connection of our outer as well as of our inner perceptions whereby they constitute one experience. . . . Though the "I," as represented through inner sense in time, and the objects in space outside me, are specifically quite distinct appearances, they are not for that reason thought as being different things.[7]

[5] Critique of Pure Reason, B409, A672 = B700.

[6] Some of the greatest of Kantian scholars have despaired of finding much meaning in the doctrine of "inner sense." Norman Kemp Smith remarks that ". . . everything which Kant wrote upon inner sense, is profoundly unsatisfactory" (A Commentary to Kant's "Critique of Pure Reason," p. 148). H. J. Paton remarks that ". . . the details of the way in which the mind affects itself, I am sorry to say, still elude me, though I have no doubt that he [Kant] is trying to say something of real importance" (Kant's Metaphysic of Experience, II, 416). Perhaps the best recent discussion of Kant's doctrine of "inner sense" is to be found in T. D. Weldon, Kant's "Critique of Pure Reason," pp. 257-270. My own treatment of the topic relies heavily, in places, on Weldon.

[7] Critique of Pure Reason, A379.

In short, in "empirical" physics, although the "outer" and the "inner" are distinct appearances, in the sense that "matter" is really given to outer sense, and the thinking "I" to inner sense, they must be regarded as "constitut[ing] one experience, " and for that reason, they must not be thought of "as being different things." In "empirical" physics, then, the distinction between "outer" and "inner" is really an analytic one which is made for purposes of separating out the various elements in experience, although in actual scientific practice, both must be regarded as intimately conjoined in a single experience.

In the second edition of the <u>Critique of Pure Reason</u>, Kant struggles to clarify his doctrine of inner sense, although in effect it remains much the same. Certain key passages that occur in the first edition are retained in the second, thus indicating their clarity in representing the author's thoughts. For instance, Kant points out that

> time is nothing but the form of inner sense, that is, of the intuition of ourselves and of our inner state. It cannot be a determination of outer appearances; it has to do neither with shape nor position, but with the relation of representations in our inner state [A33 = B49].

Hence, time and inner sense are formally quite distinct from space and outer sense. Kant, however, goes on to point out:

> Time is the formal <u>a priori</u> condition of all appearances whatsoever. Space, as the pure form of all <u>outer</u> intuition . . . serves as the <u>a priori</u> condition only of outer appearances. But since all representations, whether they have for their objects outer things or not, belong, . . . as determinations of the mind, to our inner state . . . time is an <u>a priori</u> condition of all appearances whatsoever. It is the immediate condition of inner appearances (of our souls), and thereby the mediate condition of outer appearances [A34 = B50].

Time underlies all appearances, both inner and outer, since, in order to have knowledge at all, appearances must be ordered in the mind successively.

> The apprehension of the manifold of appearance is always successive. The representations of the parts follow upon one another. . . . The appearances, in so far as they are objects of consciousness simply by virtue of being representations, are not in any way distinct from their apprehension, that is, from their reception in the synthesis of imagination; and we must therefore agree that the manifold of appearances is always generated in the mind successively [A189–A190 = B234–B235].

However, the relation between inner sense and outer sense is evidently a reciprocal one for Kant. For if, on the one hand, time can be thought of as underlying all our sensual representations, on the other hand it is equally true that space and outer sense are the primary conditions, at least in physics, for our having any self-awareness and inner sense experience at all. Kant makes this quite clear in the second edition of the Critique of Pure Reason:

[From] the existence of things outside us . . . we derive the whole material of knowledge, even for our inner sense. . . . Through inner experience I am conscious of my existence in time. . . . It is identical with the empirical consciousness of my existence, which is determinable only through relation to something which, while bound up with my existence, is outside me. . . . The reality of outer sense is thus necessarily bound up with inner sense, if experience in general is to be possible at all; that is, I am just as certainly conscious that there are things outside me, which are in relation to my sense, as I am conscious that I myself exist as determined in time. . . . [There] must therefore be an external thing distinct from all my representations, and its existence must be included in the determination of my own existence, constituting with it but a single experience such as would not take place even inwardly if it were not also at the same time, in part, outer [Bxxxixn-Bxlin].

Again, Kant points out: ". . . the representations of the outer senses constitute the proper material with which we occupy our mind" (B67). He remarks that

. . . for all inner perceptions we must derive the determination of lengths of time or of points of time from the changes which are exhibited to us in outward things, and the determinations of inner sense have therefore to be arranged as appearances in time in precisely the same manner as we arrange those of outer sense in space [B156].

Finally, Kant points out that, in order to have knowledge at all, at least in relation to the objects of outer sense, a synthesis of space and time, of the "outer" and the "inner," is necessary:

Appearances, in their formal aspect, contain an intuition in space and time, which conditions them, one and all, a priori. They cannot be apprehended, that is, taken up into empirical consciousness, save through that synthesis of the manifold whereby the representations of a determinate space or time are generated, that is, through combination of the homogeneous manifold and consciousness of its synthetic unity [B202-B203].

111

In the main, Kant has physics in mind throughout his discussion of inner and outer sense in the Critique of Pure Reason. The outcome of that discussion is that both space and time, the "outer" and the "inner," must be intimately conjoined if we are to have any knowledge of nature at all. But, if from "the existence of things outside us . . . we derive the whole material of knowledge, even for our inner sense," and if "the representations of the outer senses constitute the proper material with which we occupy our mind," what is puzzling is why inner sense is necessary at all for sense knowledge of nature. As T. D. Weldon expresses the problem: "The difficulty which at once presents itself is that the immediate data of consciousness are already exhausted [by the outer senses] and therefore there is in fact nothing left for inner sense to do."[8] All knowledge, for Kant, begins with sense experience, and the reception of such experience by the mind is always passive. Hence, inner sense must be a passive sensing; the mind is capable of affecting itself and at the same time is aware of the affection.[9] But mental activity is always an activity, and so there is an evident paradox here. Kant, however, was well aware of this paradox:

> This is a suitable place for explaining the paradox which must have been obvious to everyone in our exposition of the form of inner sense . . . namely, that this sense represents to consciousness even our own selves only as we appear to ourselves, not as we are in ourselves. For we intuit ourselves only as we are inwardly affected, and this would seem to be contradictory, since we should then have to be in a passive relation [of active affection] to ourselves.[10]

Kant goes on, in a difficult passage, to explain that what determines inner sense is the understanding and its power to combine the manifold of intuition by bringing it under an apperception. It would appear, then, that inner sense is really an awareness of awareness, an intuition of our own mental processes. Kant's teaching is quite clear concerning the fact that in order to have knowledge of nature, or physics, the manifold of intuition received through outer sense must be synthesized in the mind

---

[8]Weldon, Kant's "Critique of Pure Reason," p. 259.

[9]"Anthropologie," GS, VII, 161-162; see also Critique of Pure Reason, B67-B68.

[10]Critique of Pure Reason, B152-B153.

successively under the form of time. There can be little question about the fact that the constitution of the human mind profoundly affects knowledge of nature. But, if inner sense is an
awareness of awareness, precisely what data it adds for knowledge of nature still remains elusive. On the other hand, at
least it provides data for knowledge of the self, for even the
scientific activity we call physics reveals something of ourselves to ourselves.

However, the solution of this problem is not essential to
the central purpose, which is to discuss Kant's distinction between geography and anthropology. In separating anthropology
from geography, Kant appears to use the distinction between
outer sense and inner sense in two respects, and moreover in
respects that differ from its use in connection with "empirical"
physics:[11]

(1) When we consider anthropology cosmologically, we are
concerned with man as in the world, and as an active agent in
worldly affairs. Man is distinguished from, and elevated
above, all other creatures by virtue of possessing self-
consciousness. He is aware of himself as an "I," as distinct
from everything else outside his own consciousness. And because he possesses the notion "I," each man is a "person."[12]
But to be a person is to be something entirely natural, for personality is a natural endowment. Even genius is a "gift of nature" (p. 220). What any individual chooses to do with his
natural endowments or "gifts of nature" is, of course, another
matter. Pragmatic anthropology then, in this first respect,
is concerned with elements of self-consciousness, but with
self-consciousness as arising out of a consciousness of natural events, be these objects in the external world or natural
human endowments. Kant points out that undue introversion,
or the unhealthy preoccupation with self, can only be overcome
through a return to an interest in the external world (p. 162).

Pragmatic anthropology is contrasted with "physiological"
anthropology which is concerned with the "dark images of the
mind," with mental phenomena that operate below the level of

[11]Initially, however, it should be pointed out as obvious, that although geography is
concerned with outer sense, i.e., with objects in space that are revealed to us through
the five outer senses, geographical knowledge, just as any other form of knowledge of
nature, requires a synthesis of the manifold of outer appearances in time.

[12]Kant, "Anthropologie," GS, VII, 127.

consciousness, or with what we would today call the unconscious. Kant remarks that on the "vast map of the mind, only a few points are illuminated, " and implies that the unconscious plays an important role in our mental life (p. 136). However, as we have already seen, Kant defines physiological anthropology, in its broadest sense, as the study of "what nature makes of man. " Since geography deals with nature, and insofar as that discipline concerns itself with man, it is evidently concerned with what is natural in man, or, in some sense, with "what nature makes of man. " For instance, Kant points out that the study of the shape of human skulls "belongs more to physical geography than to pragmatic anthropology" (p. 299). Presumably, the shape of the skull is not a factor that ordinarily affects human self-consciousness, and hence is "physiological. " Further-more, there is a second respect in which geography plays a role in a general "physiological anthropology, " since place and custom form a "second nature" of which men are not conscious (p. 121). We have already seen that Kant, in his Lectures on Ethics, assigns to "moral" geography the study of customs, in the sense of unconsciously or unreflectively held mores. In another context, Kant indicates that attempts to classify whole nations as to national temperament or character, based on heredity or on customs which through long usage have become "second nature, " are a task of the empirical geographer.[13] Again, the study of how national character and climate influ-ence genius would, presumably, be the task of geography, al-though Kant does not make this clear in the context (p. 226). Perhaps a continuing and growing importance for geography, as the study of man's "second nature, " and as a contribution to a general "physiological anthropology, " can be seen in Kant's remark that the perfection of "a second nature . . . in-deed is the ultimate moral end of the human species. "[14]

To sum up: in its first respect, pragmatic anthropology deals with elements of self-consciousness, but with self-consciousness as basically conditioned by consciousness of events in the external world outside of the human mind, and with natural, physically based, human endowments. Phys-iological anthropology, as anthropology, also deals with men-

[13]"Anthropologie, " GS, VII, 312.

[14]"Conjectural Beginning of Human History, " trans. by Fackenheim, in Kant on History, ed. by Beck, p. 63.

114

tal events, but at the level of the unconscious. Geography, insofar as it is concerned with man, deals with him either in a physical and "external" sense, that is, with factors that do not ordinarily condition human self-consciousness, or with unconsciously held mores, with man's "second nature" as it conditions his behaviour unconsciously, through the peculiar exigencies of place and custom.

But the distinction between geography and anthropology, as based on the distinction between the "outer" and the "inner," evidently differs somewhat from the employment of the same distinction in connection with "empirical" physics. In the latter discipline, the distinction is only analytic, in the sense that the "inner" and the "outer" must be regarded as constituting "one experience" or "one thing"; to talk of an "outer" physics and an "inner" physics, as distinct disciplines, makes no sense. On the other hand, Kant obviously regarded it as not only possible, but necessary, to draw a clear distinction between geography and anthropology, as separate disciplines, and as based on the distinction between the "outer" and the "inner." Despite the fact that geography and anthropology together form a "whole" of our pragmatic, empirical knowledge of the world, the distinction between the "outer" and the "inner" in this context is much looser than it is in the context of "empirical" physics. The "looseness," I think, stems from the fact that there is nothing inherently necessary at the empirical level in the conditioning of man's "inner" experience by his awareness of events in the external world. In this sense Kant was no environmental determinist; the human mind, in its own mental processes, and hence in its awareness of the self, is not necessarily determined in all respects by its awareness of occurrences external to itself. Furthermore, what each man chooses to make of his natural endowments is to some extent within his own power. This "looseness" is brought out quite clearly in connection with the second respect in which Kant uses "inner sense" in anthropology. [15]

[15]The account presented in the above paragraph is somewhat oversimplified. Obviously, since geography, as a type of knowledge of nature, requires inner sense as well as outer sense, to speak of an "outer" geography and an "inner" geography, as distinct sciences, makes as little sense as it would in the case of physics. To maintain Kant's parallelism of the sciences, psychology should be set alongside physics, just as anthropology is set alongside geography. The drawback, however, is that Kant does not appear to leave room for a psychology as distinct from anthropology. As we

(2) In Part II of the Anthropologie, Kant sets himself the task of distinguishing what is "internal" in man from what is "external."[16] In this respect, anthropology as concerned with "inner sense" bears no relation whatever to "outer sense," and hence no relation to geography or physics. We are here concerned with what is essential to man's "inner" constitution, and in this respect, Kant is drawing a radical distinction between man and nature. Man is radically distinguished from nature by virtue of the possession of character. Kant points out that the concept "character" can be used in at least three distinct senses: as natural disposition, as temperament, and as mental or moral character. Although the distinction between natural disposition and temperament is not at all clear,[17] Kant remarks that these two senses of the term refer to "what

have seen, in the Critique of Pure Reason, Kant regards psychology as being merely an aspect or part of anthropology. Physics can hardly be regarded as merely an aspect of geography. In his letter of October 20, 1778, to Marcus Herz (Kant: Philosophical Correspondence, p. 91), Kant remarks that, "My empirical psychology is now briefer, since I lecture on anthropology," thus indicating that at least a part of what at one time had been labelled psychology had become anthropology. In addition, Kant was often devastatingly critical of the psychology of his day (see especially First Introduction to the Critique of Judgment, trans. by Haden, pp. 41-43). He does not, however, appear to set forth a positive doctrine of psychology to supersede what he had criticized. Nor does it appear that such a doctrine of strictly theoretical and scientific status is possible, for as Kant remarks in the context of his critique of psychology, ". . . when one seeks psychological grounds of explanation for mental events . . . so far as I know, these principles are wholly empirical with one exception, namely, that of the continuity of all changes (because time, which has only one dimension, is the formal condition of inner intuition), which is the a priori foundation of these perceptions. However, this is virtually worthless, since the general theory of time does not furnish enough material for an entire science, unlike the pure theory of space (geometry)" (p. 41). Hence, in order to be consistent, it would appear that Kant should reserve the word "psychology," as distinct from anthropology, for theories of the soul as an immaterial substance—hardly a very worthwhile science. It could perhaps be maintained that, just as physics is an abstract study of nature whereas geography is a concrete study, so psychology is an abstract study of the self whereas anthropology is a concrete study. Yet, although Kant often discusses feelings, emotions, etc., in general terms in his Anthropologie, he gives no indication here that this type of study is to be regarded as psychology. As indicated above, there are apparently, for Kant, insufficient theoretical grounds for establishing a psychology as distinct from anthropology. See Brett's History of Psychology, ed. by R. S. Peters, pp. 535-545, for an illuminating discussion of Kant's "psychology" that is based mainly on his Anthropologie. Brett adopts the point of view that psychology for Kant is merely an aspect of anthropology, and that in Kant can be found "the real beginning of 'psychology without a soul'" (i.e., of empirical and behavioural psychology), (p. 538).

[16]"Anthropologie," GS, VII, 283.

[17]Temperament presumably involves social conditioning, although Kant is never very clear about this.

116

can be done with man, " whereas moral character refers to man "as a rational creature who has acquired freedom, " and relates to "what he himself is willing to make of himself. "[18] It is character, then, in its moral sense of self-awareness, that radically distinguishes man from nature:

Here it does not matter what nature makes of man, but what man himself makes of himself, for the former belongs to the temperament (where the subject is merely passive) and the latter shows that he has a character [p. 292].

Kant goes on to remark:

That man who is conscious of a character in his mode of thinking, does not have it by nature, but must have acquired it at all times. One can also assume: that the establishment of the same resembles an act of renaissance [or rebirth]. [P. 294.]

The possession of character, in its moral sense, is something which is dependent, for its acquisition, on the action or conscious decision of an individual human being; and in this respect, it is a non-historical act. Kant points out that although an understanding of the human genus, as an animal species, can only be drawn from history, the character of the human species as human cannot be drawn from history. Human character can be understood only through the medium of human reason reflecting upon its own activity; human reason must subjectively know and modify itself individually, in order that human character, as moral character, can be possible at all (p. 413). In fact, the only empirical evidence that one possesses character is the avowal, towards oneself and also towards others, that the highest moral maxim will govern one's conduct (p. 295).

It is on the basis of the possession of moral character that Kant assigns man his place in the system of nature:

In order to assign man into the system of living nature, and thus to characterize him, no other alternative is left than this: that he has a character which he himself creates by being capable of perfecting himself after the purposes chosen by himself. Through this, he, as an animal endowed with reason (animal rationabile) can make out of himself a rational animal (animal rationale). [P. 321.]

Kant summarizes the basic message of pragmatic anthropology in the following words:

[18]"Anthropologie, " GS, VII, 285.

The sum of the pragmatic anthropology in respect to the destination of man and the characteristic of his development is as follows. Man, on account of his reason, is destined to be within human company and to cultivate, civilize, and moralize himself within it by means of art and science, no matter how great his animalistic tendency may be to resign himself passively to the in-citements of leisure and luxury which he calls the state of happiness, but rather [he is destined on account of his reason] to make himself worthy of humanity in combat with the obstacles which are affixed to him through the rawness of his nature [pp. 324-325].

Man, because of the "rawness" of his nature, is rooted in na-ture.[19] But his destiny is to overcome this "rawness," and to realize his rationality, through the development, by his own action, of good out of "evil." History is on man's side, since the possession of moral character, man's distinguishing fea-ture, "already implies a favourable natural disposition and an inclination to the good within him." On the other hand, "evil (since it holds conflict within itself and does not permit a per-manent principle) is truly without character" (p. 329). How-ever, to pursue this topic further would go well beyond a pragmatic, empirical anthropology, and into considerations of ethics and theology.[20]

GEOGRAPHY AND HISTORY

Kant leaves little doubt that he regarded history as an empirical discipline. He points out that his own philosophical speculations on history can hardly "displace the work of practicing empirical historians":

That I would want to displace the work of practicing empirical historians with this Idea of world history, which is to some extent based upon an a priori principle, would be a misinterpretation of my intention. It is only a

[19]Yet, even as a purely natural animal, man is set aside from other animals be-cause of the "marvelous dexterity" of his hands, which already indicates, according to Kant, a rational component in his makeup. "The characterization of man as a ra-tional animal rests already in the shape and organization of his hand, his fingers and finger tips, partly in their formation, partly in their fine sense of touch by which na-ture has made him skilful, not for one manner of handling goods but indefinitely for all, therefore for rational use, and hence [nature] has marked . . . his kind as a ra-tional animal" (ibid., p. 323).

[20]For a discussion of this topic, and its ethical and theological implications, see Emil L. Fackenheim, "Kant and Radical Evil," University of Toronto Quarterly, XXIII, No. 4 (1954), 339-353.

suggestion of what a philosophical mind (which would have to be well versed in history) could essay from another point of view. [21]

Furthermore, Kant indicates that his speculative philosophy of history is "no match for a history which reports the same events as an actually recorded occurrence."[22] Again, as we have already seen, in the Critique of Judgement Kant mentions history along with geography as being subject to the "faith of history"; hence, both are "rough," empirical disciplines.

Nevertheless, Kant evidently regarded history, as he did geography, as being capable of becoming, at least in part, a law-finding discipline. [23] His clearest statement on this issue is the following:

Whatever concept one may hold, from a metaphysical point of view, concerning the freedom of the will, certainly its appearances, which are human actions, like every other natural event are determined by universal laws. However obscure their causes, history, which is concerned with narrating these appearances, permits us to hope that if we attend to the play of freedom of the human will in the large, we may be able to discern a regular movement in it, and that what seems complex and chaotic in the single individual may be seen from the standpoint of the human race as a whole to be a steady and progressive though slow evolution of its original endowment. Since the free will of man has obvious influence upon marriages, births, and deaths, they seem to be subject to no rule by which the number of them could be reckoned in advance. Yet the annual tables of them in the major countries prove that they occur according to laws as stable as [those of] the unstable weather, which we likewise cannot determine in advance, but which, in the large, maintain the growth of plants, the flow of rivers, and other natural events in an unbroken, uniform course. [24]

History, as understood by Kant, is concerned with the play of human behaviour "in the large," and not with single, individual occurrences. Knowledge of the latter, presumably, is the role of "common knowledge."

Kant's main teaching nevertheless is that history, as distinct from the philosophy of history, is an empirical discipline.

---

[21]Kant, "Idea for a Universal History from a Cosmopolitan Point of View," trans. by Beck in Kant on History, p. 25.

[22]"Conjectural Beginning of Human History," trans. by Fackenheim, ibid., p. 53.

[23]For an account of the philosophical difficulties involved in the attempt to view history as a theoretical discipline, on the basis of Kantian philosophy, see Fackenheim, "Kant's Concept of History," Kant-Studien, XLVIII, No. 3 (1956-57), 381-398.

[24]"Idea for a Universal History from a Cosmopolitan Point of View," Kant on History, p. 11.

As such, it ranks with geography and anthropology in the ordering of the various sciences. Yet here, as in the case of history
conceived as a theoretical discipline, the pickings are thin, and
"the donations are grudgingly given."[25] In the introduction to
the Physische Geographie, Kant's fundamental division among
the empirical sciences, as based on the distinction between
outer sense and inner sense, is between geography and anthropology. The distinction he draws between geography and history in that work is a secondary one, since it is a distinction
that pertains solely to outer sense. Hence, in relation to geography, history is entirely the history of nature. On the other
hand, there is a history, conceived as moral history, that relates to anthropology. As Kant remarks in a late work:

> If it is asked whether the human race at large is progressing perpetually to-
> ward the better, the important thing is not the natural history of man
> (whether new races may arise in the future), but rather his moral history.[26]

Later in the same work, Kant indicates that "enthusiastic par-
ticipation in the good," as investigated in a moral history, is
important for anthropology (p. 145). Hence, it would appear
that history as an empirical discipline has no real independence,
but is subordinate to either geography or anthropology, dependent on whether the particular history is history of nature or
history of morals.

The basic concern, however, is with the relations between
geography and history, and not with those between anthropology
and history. In the introduction to the Physische Geographie,
Kant indicates quite clearly that geography is concerned essentially with space or spatial relations, whereas history is concerned with time or temporal sequences. However, since the
basic distinction Kant makes between geography and history is
one drawn within outer sense, the concept "time," as employed
in relation to the history of nature, is quite distinct from the
concept "time" understood as the form of inner sense, which
is Kant's general meaning of the term throughout the Critique
of Pure Reason, and in relation to theoretical sciences. As
employed in connection with the history of nature, the concept

---

[25]Fackenheim, "Kant's Concept of History," 384.

[26]"An Old Question Raised Again: Is the Human Race Constantly Progressing?"
trans. by Robert E. Anchor in Kant on History, p. 137.

"time" is used in an "objective" sense, rather than in the "subjective" sense of the form of inner sense. All the same, it is obvious that natural history, in order to qualify as a species of knowledge of nature, must be subject to time as the form of inner sense, and thus requires a synthesis of the manifold of outer appearances in time. (This situation presents epistemological problems for Kantian philosophy which will be remarked on later.)

Kant exhibits considerable ambivalence at times respecting the issue of clearly separating the history of nature from the "description of nature," or geography. In the introduction to the Physische Geographie, he indicates that they must be separated, since they involve fundamentally different approaches to the study of nature—the spatial organization and relations of phenomena as opposed to their change and development through time. This point of view is most strongly supported in his debate with the geographer, Georg Forster, as it appeared in his article, Über den Gebrauch teleologischer Principien in der Philosophie. [27] In that article, Kant points out that Forster's conjectures on the origin and development of the Negro race belong properly to a history of nature and not to its description. Kant goes on to argue that the separation of heterogeneous subject matter is essential to assure new scientific insights and advances. In this work, however, the history of nature is represented as being able, so far, to display only "poor fragments" and "tottering hypotheses." It may in the long run never be more than the mere "silhouette of a workable science, and for most questions the answer might be a blank" (p. 162). Kant supports this contention in the introduction to the Physische Geographie, when he points out that it is totally incorrect to believe that we have a fully-fledged history of nature. Possessing in fact only the name, we are thereby led to believe that we "possess the thing itself." "If one describes the occurrences of the whole of nature as they have been through all time, then, and only then, would one deliver a correct history of nature, as it is called." In Kant's day, as indeed in our own, no such exact science of the history of nature existed.

Nevertheless, Kant was not opposed, in principle, to a history of nature. As he remarks on one occasion:

[27] GS, VIII, especially pp. 161-163.

> Natural history is as yet far from sufficiently developed to furnish the reasons for the manifold diversity of . . . deviations. However much one may hate the impertinence of mere opinions—and justly so—one must dare a History of Nature which is a distinct science, and which gradually could progress from opinions to discernment. [28]

Kant lived before the scientific development of historical geology and the theory of evolution. Nevertheless, there can be little question that he was correct in criticizing the speculations of Linnaeus, Buffon, and others on the history of nature as being not scientifically respectable, and as lacking sufficient empirical evidence.

In other contexts, as noted earlier, Kant displays much less assurance in drawing a sharp distinction between the history of nature and the description of nature. This lack of assurance appears to stem in part from the fact that the term "history," as Kant himself points out, was still being very widely used in his day in the sense of the Greek term "historia," which meant both a history and a description of nature. [29] Kant's apparent lack of certainty, especially in his less polemical works, probably stems from this concession to current usage. For instance, in one context in which he fails to draw the distinction sharply, he remarks in a footnote:

> We commonly take the meaning of the description of nature and the history of nature in the same sense. However, it is clear that knowledge of the things of nature as they are at present makes desirable the knowledge of that state in which these things have once been, and of the sequence of changes through which they have gone so that they can be in their present state as they now are. The natural history which we are lacking almost completely would teach us the transformations of the form of the earth, at the same time as the transformations of the creatures of the earth (plants and animals). . . . This discipline would probably lead a great quantity of seemingly diverse species back to the very same kind of race, and would transform the present loose school system of natural description into a physical system fit for the understanding. [30]

What is interesting in this passage is that although Kant draws the distinction between a "history of nature" and a "description of nature," there is evidently an interdependence between the two, since an adequate understanding of the present "descrip-

[28]On the Different Races of Man, p. 24.

[29]"Über den Gebrauch teleologischer Principien in der Philosophie," GS, VIII, 162.

[30]"Von den verschiedenen Racen der Menschen," GS, II, 434n.

tion of nature" depends upon the development of a scientific history of nature. In short, Kant is raising the issue of the
explanatory dependence of geography on historical knowledge.
Kant, of course, was strongly opposed to what he called the
"scholastic" or "customary" description of nature. This approach to geography was inadequate on two counts: (1) it was a
mere aggregate of information and hence failed to qualify as
systematic, scientific knowledge; and (2) it bore no relation
to the requirements of life or knowledge, and hence failed as
a propaedeutic. It was Kant's hope that geography, as conceived by him, and understood as a systematic yet empirical
science, would overcome these deficiencies in the inadequate
conception of the subject current in his day.

In later years, Kant apparently despaired of the acceptance
of a clear distinction between a history of nature and a description of nature, for he remarks on one occasion:

If the name of natural history, now that it has once been adopted, is to continue to be used for the description of nature, we may give the name of archaeology of nature . . . to that which the former literally indicates, namely
an account of the bygone or ancient state of the earth—a matter on which,
though we may dare not hope for any certainty, we have good ground for con-
jecture. Fossil remains would be objects for the archaeology of nature.
. . . For, as work is actually being done in this department, under the name
of a theory of the earth, steadily though, as we might expect, slowly, this
name would not be given to a purely imaginary study of nature, but to one to
which nature itself invites and summons us. [31]

From the broader context in which this passage occurs, it is
evident that the concept "archaeology of nature" refers to both
"ancient" states of the earth, and the transformations it has
undergone through time. And since natural history is to be
understood in this quotation in the sense of a description of nature, Kant obviously intends a threefold division: "ancient"
states of the earth, transformations through which the earth
and its creatures have gone, and the present state of the earth
and its inhabitants. The previous quotation also clearly indicates this threefold division. Furthermore, the same distinctions are intended in the introduction to the Physische Geographie,
although in that context the matter is perhaps not so clearly put.
Relevant passages include the following:

[31]Critique of Judgement, Part II, sec. 82, p. 90n.

History is description according to time; description according to space is geography.

History and geography enlarge our knowledge with respect to time and space. History concerns events which, under the aspect of time, have occurred one after the other. Geography concerns appearances under the aspect of space which occur simultaneously. . . .

The history of what occurs at various times is history proper, and is nothing else than a continuous geography. Therefore, it is a great historical lack when one does not know in which location the events took place, or what the conditions were.

History therefore differs from geography only in respect to space and time. The former, as mentioned above, is an account of occurrences which have succeeded each other, and relates to time. The latter, however, is an account of occurrences which take place beside each other in space. History is a narrative, but geography is a description. Therefore we may have a description of nature, but not a history of nature. . . .

The history of nature contains the manifold qualities of geography, namely how things were in different epochs, but not how things are at the present time because this would be a description of nature. . . .

The name geography therefore designates a description of nature, and at that of the whole earth. Geography and history fill up the total span of our knowledge; geography namely that of space, but history that of time.

We ordinarily assume that there is an old and a new geography, because geography has existed at all times. But which came first, history or geography? The latter is the foundation of the former, because occurrences have to refer to something. History is in never relenting process, but things change as well and result at times in a totally different geography. Geography therefore is the substratum. Since we have an ancient history, so naturally we have an ancient geography.

We know best the geography of the present time. The present geography serves, among other ends, to elucidate the ancient geography by using the ancient history. . . .

Let us attempt a brief summary of the essentials of Kant's argument, which appears to be somewhat disjointed probably as the result of its having been filtered through undergraduate lecture notes. The surface of the earth is the substratum or the stage on which the events of history take place. Hence, history always requires a basic knowledge of geography. However, historical knowledge is required in geography in two senses: (1) we require knowledge of changes that have occurred in the past in order to understand the present state of the earth; and (2) we require historical knowledge to reconstruct the past states of the earth, or for "ancient" geography. In short, geography is ontologically prior to history, although history is epistemologically prior to geography. However, when Kant remarks that "the history of what occurs at various times is history proper, and is nothing else than a continuous geography" what he means is far from clear. To begin with, the quotation illustrates the ontological priority of geography, and the subordination of the history of nature to geography. But it can hardly be taken to mean literally that history in itself or "history proper" is nothing more than a "continuous geography," since Kant has clearly indicated, in several contexts, that the perspectives, or approaches to nature, of a history of nature and a description of nature are different, and must therefore be separated, even if significant interdependencies remain between them. The clue is probably contained in the phrase "at various times." In other words, we can, so to speak, stop the clock at various points in time, and attempt to reconstruct, on the basis of historical evidence, what the geography, or spatial relations among phenomena on the earth's surface, would probably have been at those various times. In short, any number of "ancient" geographies are possible.

The epistemological problem, referred to earlier, stems from the fact that in order to have knowledge all outer appearances must be synthesized by the mind under the form of time. But since time, as the form of inner sense, can refer only to the sense experience of each separate individual, then precise scientific knowledge is only possible on the basis of individual sense experience. However, the concept "time" as employed in connection with the history of nature has only an "outer" and not an "inner" denotation. How, then, can we have any systematic, scientific knowledge of the remote history of the earth and its inhabitants? Somehow these remote events would have to be capable of being synthesized under the form of time, in order to render a strictly scientific history of nature possible. Kant discusses the problem in the following passage:

. . . we can say that the real things of past time are given in the transcendental object of experience; but they are objects for me and real in past time only in so far as I represent to myself (either by the light of history or by the guiding-clues of causes and effects) that a regressive series of possible perceptions in accordance with empirical laws, in a word, that the course of the world, conducts us to a past time-series as condition of the present time—a series which, however, can be represented as actual not in itself but only in the connection of a possible experience. Accordingly, all events which have taken place in the immense periods that have preceded my own existence mean really nothing but the possibility of extending the chain of experience from the present perception back to the conditions which determine the perception in respect of time.[32]

The "objects" of past time are "transcendental object[s] of experience," that is, they are not given to us in our immediate sense experience, but are postulated by human reason. We are aware, through our own present experience of the world, that things change; hence, it is reasonable to postulate that, in the time before each of us was present on the earth, things were different than they now are, and certain events took place which have been responsible for the present condition of the earth. Although we can never have direct knowledge of these events, it is possible to gain some insight into the past through two means: (1) "the light of history"; and (2) chains of causal reasoning that proceed back in time from what we can experience at present. It seems that what Kant means by the "light of history" are objects from the past which can actually be present to our senses now. These would include fossil remains, human artifacts, and written documents from the past. In addition, since geography and history are "rough," empirical disciplines, the oral testimony of others would also be included as historical evidence. However, as Kant points out in the introduction to the Physische Geographie, written documents are always preferable to, and more reliable than, material that has been handed down through an oral tradition.

Kant sheds further light on the issue in the following passage:

Herr Forster thinks a history of nature, regarding, for instance, the first origin of plants and animals, is not a science for human beings, but for gods who were present or themselves the authors of those events. However, the history of nature, as I see it, would merely explore the connexion between the present qualities of natural objects and their causes in earlier times— guided by laws which we do not invent but derive from the observable natural forces, tracing them back as far as analogy permits.[33]

---

[32]Critique of Pure Reason, A495 = B523.

[33]"Über den Gebrauch teleologischer Principien in der Philosophie," GS, VIII, 161-162; trans. by Rabel, Kant, p. 185.

Investigations into the history of nature begin with what is present to our senses here and now, and proceed back in time to ever remoter causes. What is present to our senses here and now would presumably include historical data as revealed by the "light of history." Such investigations proceed through the employment of empirical laws as these have been derived from the study of the forces of nature in the present. But we cannot apply these laws directly, since we have few or no data to which to apply them. Rather, historical investigation proceeds by analogy and hypothesis. These procedures are peculiar to empirical sciences, and in no way apply to mathematical or theoretical sciences, since the latter give us apodictic certainty and hence do not require reasoning by analogy and hypothesis.

Kant points out that "in the case of two dissimilar things we may admittedly form some conception of one of them by an analogy which it bears to the other." He goes on to define analogy in the following terms:

Analogy . . . is the identity of the relation subsisting between grounds and consequences—causes and effects—so far as such identity subsists despite the specific difference of the things, or of those properties, considered in themselves (i.e., apart from this relation), which are the source of similar consequences.

Perhaps one of Kant's illustrations will clarify the matter:

Thus on the analogy of the law of the equality of action and reaction in the mutual attraction and repulsion of bodies I am able to picture to my mind the social relations of the members of a commonwealth regulated by civil laws; but I cannot transfer to these relations the former specific modes, that is, physical attraction and repulsion, and ascribe them to the citizens, so as to constitute a system called a state. [34]

Kant appears to use the term "analogy" in much the same sense as the term "model" is employed by some contemporary philosophers of science. Certain structural similarities may exist between otherwise disparate branches of knowledge. By employing concepts drawn from a field of research that is well-developed, and by applying them to a field about which we know relatively little, some important insights into the latter field

[34] Critique of Judgement, Part II, sec. 90, pp. 136-137.

may be gained. Hence, analogies or models serve basically as heuristic devices. But, as Kant's example clearly illustrates, despite the fact that we may profitably think about aspects of human society on the model of the attraction and repulsion of bodies in physics, we are in no sense entitled to impute to social relations the specific modes of attraction and repulsion of bodies as understood in physics.

But how, specifically, are analogies applied to historical studies? Apart from the very loose sense of thinking about the origins of human history on the basis of an "analogy" with the Book of Genesis, in his Conjectural Beginning of Human History, Kant nowhere appears to discuss this matter, nor does he ever discuss it in connection with geography, although he clearly indicates that the employment of analogies is a procedure in historical research, and in empirical sciences generally. Kant might have sanctioned the employment of analogies in two senses. (1) In reasoning back to remoter causes from our present knowledge of the state of the earth, in cases where certain effects are in evidence which bear some resemblance to certain contemporary effects of which we have causal knowledge, we could reason, by analogy, that similar effects will be produced by similar causes. (2) In cases where we possess isolated evidence from the past, e.g., fossil remains or human artifacts, this evidence could be regarded as effects, and if these effects bear some resemblance to certain contemporary effects of which we have causal knowledge, again it could be reasoned that similar effects will be produced by similar causes. At best, however, such procedures would give us only possible insights into historical processes, and not apodictic certainty.

Hypotheses differ from analogies in that in the case of the former we are not applying what is known in one field of knowledge to elucidate knowledge in another field, but rather we are reasoning back abductively from known effects to possible causes without employing concepts drawn from what is already known.

We remarked above with reference to probability that it is only an approximation to certainty. Now this is especially the case with hypotheses, by means of which we can never attain to apodictic certainty, but only to a degree of probability sometimes greater, sometimes less.

A Hypothesis is an assent of the judgment to the truth of a principle on account of the sufficiency of the consequences. . . .

> All assent, then, in the case of hypotheses is founded on this, that the supposition taken as a principle is adequate to explain other cognitions as consequences from it. . . .
>
> Yet in every hypothesis there must be something apodictically certain, viz.: First—<u>The possibility of the supposition itself.</u>

As an example of an acceptable hypothesis, Kant cites the following:

> For instance, when we assume a subterranean fire for the purpose of explaining earthquakes and volcanoes, such a fire must be possible, if not as a flaming, yet as a hot body.

Kant proceeds to give an example of an unacceptable hypothesis:

> But for the sake of explaining certain other phenomena, to make the earth an animal in which the circulation of the internal juices produces heat, is to put forward a mere fiction, and not a hypothesis. [35]

Kant could have pointed out, although he does not, that the acceptable hypothesis is strengthened by analogy with our direct knowledge of hot bodies and the effects they produce. On the other hand, in the case of the unacceptable hypothesis, we have no reason whatsoever to believe that the earth can profitably be regarded as an animal body, and that the interior of the earth functions on the model of the interior of an animal. In this case, we are not putting forth an analogy or model, but a "mere fiction." Presumably, in Kant's view, any hypothesis that is put forward to explain some aspect of the history of nature will be strengthened if it can be supported by an analogy drawn from contemporary knowledge of the present state of the earth. This presumption, in turn, rests upon an assumption respecting the uniformity of nature.

In conclusion, it would appear that in Kant's considered opinion the history of nature would never be more than an empirical science, dependent upon the employment of hypotheses and analogies for whatever insights it could furnish. Although at times Kant entertained the possibility that the history of nature was capable, in principle, of becoming a theoretical discipline, this eventually would not appear to be possible on the basis of Kantian philosophy. Since so much of the evidence of history is isolated, scattered, discontinuous, or missing, and since we

[35]<u>Introduction to Logic</u>, pp. 75-76.

can often not get direct perceptual access to the evidence, much of it would remain incapable of being synthesized in the mind under the form of time. Hence, the history of nature appears doomed to remain at the empirical level. It can provide us with useful insights and varying degrees of probable knowledge, but not with strictly scientific knowledge. And, insofar as geography depends on a precise science of history for its explanatory validity, it too is permanently relegated to the empirical level.

In addition, epistemological problems undoubtedly also arise in connection with the concept "space," the basic concept of geography, although the issue is never discussed by Kant, as it is in connection with the concept "time." In the case of "ancient" geographies, where a reconstruction of past spatial states of the earth is being achieved, since the reconstruction evidently involves material that could never have been the immediate perceptual experience of any one individual, the concept "space" cannot strictly mean "the form of outer sense" as it does in connection with physics and theoretical sciences generally. Even in the case of a present "description of nature" of the whole earth, since it can only be achieved through a compilation of the testimony of many observers, and hence involves a construction from the experiences of many people, again the concept "space" cannot strictly mean what it does in physics. It would appear, then, that in connection with the empirical sciences, geography and history, the concepts "space" and "time" acquire additional "objective" meanings which they do not possess when employed in connection with theoretical science.

APPENDIX

Speaking very generally, there are two alternative interpretations of Kant's doctrine of space and time in relation to the empirical sciences, geography and history. The first alternative interpretation would maintain that if geography and history are to qualify as sciences, and if they are to serve as a propaedeutic for more advanced work, then the doctrine of space and time as it relates to the empirical sciences must be substantially the same doctrine that relates to the theoretical sciences. But even if one grants that Kant has solved the problem of intersubjectivity, i.e., that his extremely subjective basic definition of time as "nothing but the form of inner sense" is reconcilable with his "objective" account of time, i.e., that all objectively discernible scientific temporal experiences are subject to the category of causality and hence involve the observer in an irreversible temporal sequence of events (Critique of Pure Reason, A189 ff. = B232 ff.), although this reconcilability has been denied by some commentators (see Norman Kemp Smith, A Commentary to Kant's "Critique of Pure Reason," p. 137), it is difficult to see how causality and the reconstruction of events are strictly speaking compatible. In other words, geography and history as sciences depend basically upon a reconstruction of events out of the personal experiences of a number of observers. This reconstruction is not reducible to causality, i.e., it is not subsumable under the category of causality, and hence cannot be explained solely in terms of causality. It is on the basis of this fact of reconstruction that I have maintained that space and time for Kant in relation to the empirical sciences take on additional "objective" meanings which they do not possess when employed in relation to theoretical science.

The second alternative interpretation is that of Kaminski (Über Immanuel Kants Schriften zur physischen Geographie, p. 72), who maintains that Kant conceives space and time in relation to geography and history in a totally different sense than he conceives them in the context of the Critique of Pure Reason, i.e., he conceives them as purely objective concepts and not as the forms of outer and inner sense respectively. In my opinion, however, this interpretation is untenable, since Kant obviously intends that geography and history qualify as sciences in at least a minimal sense of that term, and since in order to qualify even minimally as sciences the phenomena with which they deal must be subject initially to space and time as the forms of outer and inner sense. Moreover, for Kant, even elementary experiences themselves, in order to qualify as minimal scientific experiences, and hence in order to be capable of being related to other experiences, must be subject basically to space and time as the forms of outer and inner sense.

VI
Kant's concept of geography
(3) Pragmatic and teleological
aspects of geography,
and the place of geography in
a classification of the sciences

GEOGRAPHY AS PROPAEDEUTIC AND
AS END-PRODUCT OF KNOWLEDGE

A number of quotations have already been given that illustrate geography's role as a propaedeutic. The essentials of Kant's position on the subject can be briefly summarized.

Geography serves as a propaedeutic to both science and life. As a propaedeutic to science, geography's role begins with the education of the young child:

The first lessons in science will most advantageously be directed to the study of geography, mathematical as well as physical. Tales of travel, illustrated by pictures and maps, will lead on to political geography. From the present condition of the earth's surface we go back to its earlier condition, and this leads us to ancient geography, ancient history, and so on.[1]

[1]Kant, Education (1803), sec. 70, p. 75.

It is map-work, however, that initially introduces the child to the discipline of scientific endeavour. Map-work is capable of doing this because maps are not only fascinating, and thereby appeal to the imagination, but at the same time they impose limitations on the free play of the child's imagination.

Children generally have a very lively imagination, which does not need to be expanded or made more intense by the reading of fairy tales. It needs rather to be curbed and brought under rule, but at the same time should not be left quite unoccupied. There is something in maps which attracts everybody, even the smallest children. When they are tired of everything else, they will still learn something by means of maps. And this is a good amusement for children, for here their imagination is not allowed to rove, since it must, as it were, confine itself to certain figures. We might really begin with geography in teaching children [sec. 73, p. 78].

Kant goes on to remark, in a passage reminiscent of Rousseau's approach to the education of children, that "we understand a map best when we are able to draw it out for ourselves" (sec. 75, p. 80). In addition, geography is of importance in the education of the child since it lays stress on location, and thereby affords exercise in "the capability of recalling the exact position of places where we have seen certain things" (sec. 60, p. 62).

What is most striking and most original in Kant's approach to geography as an academic discipline, however, is the importance he attaches to its role in university education. As we have already seen, he set forth geography's role most clearly in the announcement of his courses for the academic year 1765-66.[2] It was Kant's conviction that most university instruction suffered from the fact "that the youth early became versed in logical subtleties without having sufficient historical [i.e., empirical] knowledge" (p. 305). Hence, he prefaced his lectures on metaphysics with empirical information drawn from physics and psychology, and those on ethics with empirical information about man and actual moral situations. Geography's role as a propaedeutic in relation to science appears to be twofold: (1) since geography studies some of the same objects as physics, it then serves as an elementary introduction to physics and other theoretical natural sciences; but (2) it evidently played a much broader role than did any other propaedeutic, since it provided a "general framework for knowledge" in which

[2]"Nachricht von der Einrichtung seiner Vorlesungen in dem Winterhalbenjahre von 1765-1766, " GS, II, 303-313.

all our empirical knowledge of the world could find a place. However, it does not follow that physics can be derived from geography, any more than ethics can be derived from anthropology.

> . . . just as a metaphysic of nature must also contain principles for applying those universal first principles of nature as such to objects of experience, so a metaphysic of morals cannot dispense with principles of application; and we shall often have to take as our object the particular na- ture of man, which is known only by experience, to show in it the implications of the universal moral principles. But by this we in no way detract from the purity of these principles or cast doubt on their a priori source. That is to say, in effect, that a metaphysic of morals cannot be based on anthropology but can be applied to it.[3]

The empirical and the theoretical sciences, in their relations with each other, do not form an hierarchical order, in that we cannot pass upwards from the empirical level to the theoretical. Nevertheless, general theoretical principles must be applicable at the empirical level, if science is to be possible and if life is to be lived rationally. Man does not live in a vacuum but in the world. Hence, in order to know under what specific sets of circumstances general theoretical principles can be applied, empirical knowledge of the world is necessary. In this sense, wide geographical and anthropological knowledge provide an antidote to the tendency to become preoccupied with logical subtleties that may bear no relation whatever to the world of experience.

This discussion leads into geography's role as a propaedeutic to life. In its association with anthropology, geography provides knowledge of the surface of the earth, which is the "home of man," or the stage upon which man will act out his life. Hence, a preliminary knowledge of man's environment is necessary for judicious living. The more a man knows about the world, the better equipped he is to live in it and cope with it. In a narrower and more specific sense, however, as Kant points out in the introduction to the Physische Geographie, a knowledge of geography is necessary for political life. Without such knowledge a man cannot read the news of the day with

---

[3]Kant, "The Doctrine of Virtue," Part II of The Metaphysic of Morals (1797), trans. by Mary J. Gregor, p. 14.

understanding.[4] In order to carry out his responsibilities as a citizen, and make intelligent decisions on matters of state and foreign affairs, a man must be well informed about the geography, not only of his own country but of other nations. Presumably geography serves also as a propaedeutic to economic life, since without specific empirical knowledge of economic conditions, and the physical basis and requirements of various crops and products, business would not be possible.

Besides being a propaedeutic, geography serves also as an end-product of knowledge. Although Kant was concerned to some extent in the 1760's with geography as a research discipline, in later years he became preoccupied with the subject as an academic discipline. As he remarks in the introduction to the Physische Geographie, physical geography is a "general compendium of nature," and one of its major functions is that it "provides a purposeful arrangement of our knowledge" of the world. In this respect we may contrast geography with "empirical" physics. In the latter, "nature is the complex of all the objects of experience," whereas in the former, nature or the world is the "sum-total of all knowledge of experience." Hence, in geography conceived as an end-product, emphasis is laid on "knowledge of experience" rather than on "objects of experience." Geography, then, as an academic discipline, is synoptic, and attempts to combine, in systematic fashion, the findings of more specialized and precise disciplines. Obviously though, one could not hope to encompass "all knowledge of experience" within the confines of a single academic discipline. Thus, geography and anthropology, in their academic roles, were, for Kant, "popular" disciplines, but popular in the best sense of that term, combining both scholarship and easily communicable information.

---

[4]It is interesting in this respect to observe that a prominent contemporary geographer has remarked that "The ideal geographer should be able to do two things: he should be able to read his newspaper with understanding, and he should be able to take his country walk—or maybe his town walk—with interest. That is all, no more, no less" (Darby, The Theory and Practice of Geography, p. 20). The idea, however, is apparently not original with Kant. Baker points out that John Locke had declared "without a knowledge of geography gentlemen could not understand a newspaper" (The History of Geography, p. 98). The remark, wherein Locke uses "Gazette" instead of "newspaper," occurs in his essay, "Some Thoughts Concerning Reading and Study for a Gentleman," The Works of John Locke (London, 1823), III, 291-300; the remark is on p. 298.

To avoid pedantry, there is required not only an extensive acquaintance with the sciences themselves, but also with the use of them. Hence it is that only the truly learned man can free himself from pedantry, which is always the attribute of a shallow mind.

In endeavouring to give our knowledge the perfection of scholastic thoroughness, and at the same time popularity . . . we must first of all look to the scholastic perfection of our knowledge . . . and only after that study how to make this methodically learned knowledge truly popular, that is, easy and generally communicable to others, yet so that thoroughness is not supplanted by popularity. . . .

. . . this truly popular perfection is in reality a great and rare perfection, which gives evidence of much insight into science.[5]

GEOGRAPHY AND TELEOLOGY

Throughout the Physische Geographie, Kant does not discuss geography in relation to teleology. In other contexts, however, he does discuss teleology in relation to material that is clearly geographical.[6] At first glance, it is perhaps puzzling that Kant does not mention teleological principles in the Physische Geographie, since one of its primary purposes was to present "a purposeful arrangement of our knowledge" of the world. However, "purposeful arrangement" in this context is being used in a purely pragmatic sense, to indicate geography's role as a propaedeutic to both science and life. Moreover, to introduce teleology into geography would be to introduce a further non-empirical or a priori element into that discipline. Nevertheless, when considered as a research discipline, rather than as a purely academic discipline, teleology evidently has a role to play in geography. Yet, teleological principles or judg-

---

[5]Kant, Introduction to Logic, pp. 37-38.

[6]See, for instance, Critique of Judgement, Part II, sec. 63, "Of the Relative as Distinguished from the Inner Purposiveness of Nature, " and sec. 82, "Of the Teleological System in the External Relations of Organized Beings"; and Kant, Perpetual Peace, "First Supplement: Of the Guarantee of Perpetual Peace, " trans. by Beck, pp. 24-32. In these contexts, Kant discusses, among other matters, the fact that rivers bring down rich alluvium that is suitable for the growth of plants which in turn provide food for mankind; that sandy soil as deposited in northern regions is well suited for the growth of pine trees; that snow in cold countries protects crops from the frost and also favours human intercourse by means of sleighs; that drift-wood is carried by the Gulf Stream to northern tundra climates, thus facilitating human existence.

ments are neither theoretical, empirical, nor practical, since they add nothing specific to either our theoretical or empirical knowledge of nature, or to our understanding of human freedom:

Now this transcendental concept of the finality of nature is neither a concept of nature nor of freedom, since it attributes nothing at all to the Object, i.e., to nature, but only represents the unique mode in which we must proceed in our reflection upon the objects of nature with a view to getting a thoroughly interconnected whole of experience, and so is a subjective principle, i.e., maxim, of judgement. For this reason, too, just as if it were a lucky chance that favoured us, we are rejoiced . . . where we meet with such systematic unity under merely empirical laws: although we must necessarily assume the presence of such a unity, apart from any ability on our part to apprehend or prove its existence.[7]

Men have a deep-rooted and basic need to regard not only their own lives but also the whole of nature as forming a purposive system. Hence, man must assume "the presence of such a unity." But it is only a "subjective" principle, an expression of the limitations of finite human comprehension. Man is so constituted by his finiteness that he cannot possibly see the whole course of nature laid out before him. Consequently, of necessity, he must impute a purpose, not only to his own existence, but to the whole course of nature as well. This teleological judgment which man passes upon the course of nature, however, is not an aspect of knowledge, because it cannot be understood or proved by recourse to experience or causal explanation. But by the mere understanding of a mechanical course of events as a necessary causal sequence, the purpose which any phenomenon or course of events is to fulfill will never be revealed to him. This is especially so regarding living organisms:

It is, I mean, quite certain that we can never get a sufficient knowledge of organized beings and their inner possibility, much less get an explanation of them, by looking merely to mechanical principles of nature. Indeed, so certain is it, that we may confidently assert that it is absurd for men even to entertain any thought of so doing or to hope that maybe another Newton may some day arise, to make intelligible to us even the genesis of but a blade of grass from natural laws that no design has ordered. Such insight we must absolutely deny to mankind.[8]

[7]Kant, Critique of Judgement, "Introduction," pp. 23-24.

[8]Ibid., Part II, sec. 75, p. 54. This idea, however, is by no means a late one for Kant. For instance, he made substantially the same point in 1755 in his General Natural History and Theory of the Heavens in Kant's Cosmogony, p. 29.

The imputing of purpose to nature is necessary in order that man may see natural events and the course of nature as an integrated whole. Yet, despite the fact we have good grounds for believing that man "is the ultimate purpose of creation here on earth," the "devastations of nature," such as floods, earthquakes, and inundations of the sea, which are often detrimental to man, would seem to indicate that he is often not comprehended in the "purposes of nature." This antinomy merely reinforces the "subjective" quality of teleological judgments.[9]

Because of the "subjective" status of teleological principles, they can in no sense replace a theoretical understanding of nature. As Kant points out, where we possess an adequate theoretical understanding of nature, teleological principles are no explanatory substitute. They are useful in natural science only when theory is lacking, although teleology can never supply a theory.[10] Hence, teleological principles, in relation to science, serve basically as heuristic devices. In relation to geography as an empirical science, teleological principles would therefore open up certain avenues of approach, giving us possible leads towards an understanding of geographical phenomena. For instance, in discussing the shape of the earth in relation to mountains and seas, Kant indicates that a teleological principle respecting providential design may enable us to make a number of discoveries. Yet, he points out that

. . . provided we restrict ourselves to a merely regulative use of this principle, even error cannot do us any serious harm. For the worst that can happen would be that where we expected a teleological connection, we find only a mechanical or physical connection. In such a case, we merely fail to find the additional unity; we do not destroy the unity upon which reason insists in its empirical employment.[11]

Moreover, even if we assume the truth of a providential design of the earth, "yet, wise as this arrangement is; we feel no scruples in explaining it from the equilibrium of the formerly fluid mass of the earth" (A687n = B715n). In another context where he discusses the apparent design of nature, and its "truly

[9]Kant, Critique of Judgement, Part II, sec. 82.

[10]Kant, "Über den Gebrauch teleologischer Principien in der Philosophie," GS, VIII, 159.

[11]Critique of Pure Reason, A687-A688 = B715-B716.

marvellous assemblage of many relations, " which enables men to live in inhospitable regions of the far north because, for instance, of the presence of moss which feeds reindeer which in turn provide for a number of man's basic needs, Kant concludes his discussion by remarking:

On the contrary it would seem audacious and inconsiderate on our part even to ask for such a capacity, or demand such an end from nature—for nothing but the greatest want of social unity in mankind could have dispersed men into such inhospitable regions. [12]

In other words, although we may think of nature as designed, even in inhospitable northern regions, to enable men to eke out a bare living, we can in fact account for man's presence in such regions on purely historical and empirical grounds, as being due to his having been forced into such regions of the earth by more powerful peoples. But the regarding of even northern regions as designed might lead us to discover a number of relations or associations among phenomena peculiar to such regions.

In a different, yet more specific and empirical sense, it would appear that teleological judgments as reflective play a possible role in geography, since in reflective judgments "only the particular is given and the universal has to be found for it" ("Introduction, " p. 18).

This judgement, also, is equipped with an a priori principle for the possibility of nature, but only in a subjective respect. By means of this it prescribes a law, not to nature . . . but to itself . . . to guide its reflection upon nature. This law may be called the law of the specification of nature in respect of its empirical laws. It is not one cognized a priori in nature, but judgement adopts it in the interests of a natural order, cognizable by our understanding, in the division which it makes of nature's universal laws when it seeks to subordinate to them a variety of particular laws ["Introduction, " p. 25].

This is a difficult passage; however, what Kant appears to be saying is that we cannot deduce the most specific empirical laws from the a priori principles of nature in general, and hence cannot regard them from the standpoint of the understanding as forming a systematic unity, but rather we require an additional a priori principle, a reflective teleological one, in order to see nature at the strictly empirical level as a "natural order" or

[12]Critique of Judgement, Part II, sec. 63, p. 16.

"system of nature." Kant expands on this theme in considerably more detail in the First Introduction to the Critique of Judgment. In that work he points out:

> For unity of nature in space and time and unity of our possible experience are one and the same, because the former is a sum of mere appearances . . . which can possess its objective reality only in experience, which itself must be possible as a system under empirical laws if (as one must) one thinks it as a system. Therefore it is a subjectively necessary, transcendental presupposition that this dismaying, unlimited diversity of empirical laws and this heterogeneity of natural forms does not belong in nature, that, instead, nature is fitted for experience as an empirical system through the affinity of particular laws under more general ones.
>
> This presupposition is, then, the transcendental principle of the faculty of judgment. [13]

Kant is here concerned with nature as a purely "empirical system." But, since its systematic character is not deducible from the a priori principles of a nature in general, in order that nature as an "empirical system" is to be possible at all, that is, in order that it does not degenerate into total particularity and heterogeneity, and hence become incomprehensible to man, a special a priori principle, a reflective principle of judgment, is necessary. Kant enlarges on this point in several later passages in the same work:

> Neither understanding nor reason can establish a priori such a law of nature. For though we can perhaps see that nature submits to our understanding in its purely formal laws (by which it is the object of experience in general), in respect to the plurality and heterogeneity of the special laws it is free from all restrictions legislated by our cognitive faculty; it is a sheer assumption on the part of the judgment for its own use in ascending continuously from particular empirical laws to more general, though still empirical ones, in order to consolidate empirical laws, which establishes that principle. Under no circumstances can a principle like this be posited to the account of experience, because only under this assumption is it possible to order experience in a systematic fashion [pp. 15-16].

> Now to be sure, the pure understanding (through synthetic principles) teaches us to regard everything in nature as part of a transcendental system through a priori concepts (the Categories); but the judgment, which as reflective is also in search of concepts for purely empirical representations, must for this purpose assume in addition that nature in its unlimited plurality has hit upon a division into genera and species which enables our judgment to find

---

[13]Trans. by Haden, pp. 14-15.

140

harmony when comparing natural forms, and to reach empirical concepts and their interconnection by ascending to more universal, yet empirical concepts; i.e., the judgment posits a priori, and hence by a transcendental principle, a system of nature under empirical laws as well [p. 17n].

The reflective judgment thus works with given appearances so as to bring them under empirical concepts of determinate natural things not schematically, but technically, not just mechanically, like a tool controlled by the understanding and the senses, but artistically, according to the universal but at the same time undefined principle of a purposive, systematic ordering of nature. Our judgment is favored, as it were, by nature in the conformity of the particular natural laws (about which the understanding is silent) to the possibility of experience as a system, which is a presupposition without which we have no hope of finding our way in the labyrinth of the multiplicity of possible special laws [p. 18].

In what may be the most important single passage in this context for our purposes, Kant goes on to remark:

Now it is clear that the nature of the reflective judgment is such that it cannot undertake to classify the whole of nature by its empirical differentiation unless it assumes that nature itself specifies its transcendental laws by some principle. This principle can be none other than that of conformity to the power of judgment itself, finding in the infinite multiplicity of things subject to possible empirical laws enough kinship to bring them under empirical concepts (classes) and these under more universal laws (higher genera), and thus to achieve an empirical system of nature. Since this kind of classification is not ordinary experiential knowledge, but is rather an artistic knowledge, insofar as nature is thought in such a way that it can be rendered specific by this kind of principle, it is regarded as art. The judgment thus necessarily carries in itself an a priori principle of the technic of nature, which is distinguished from the nomothetic of nature through transcendental laws of the understanding in that the latter can validate its principle as law, but the former only as a necessary presupposition [p. 20].

H. W. Cassirer,[14] in his commentary on the First Introduction to the Critique of Judgment, points out that Kant is doing something repecting nature that is very different from anything he does in the Critique of Pure Reason. In the latter, he is concerned primarily with demonstrating the dependence of the possibility of scientific experience, and hence of science, on the universal a priori laws of the understanding. In the former, however, he is concerned with an "empirical system of nature," with the arrangement of all particular empirical laws of nature into a system. In this context, then, Kant is no longer concerned

<hr />

[14]A Commentary on Kant's "Critique of Judgment," especially pp. 108-113.

with the necessary conformity of empirical laws to universal a priori laws of reason (this can be taken for granted), but with the further problem, which is confined to the strictly empirical level, of creating an "empirical system of nature, " so that specific or particular empirical laws can be exhibited as connected with one another. However, I cannot agree entirely with Cassirer when he remarks that, "It is plain that Kant is here concerned with an entirely new problem, a problem of the existence of which he knew nothing when he wrote the Critique of Pure Reason and the Prolegomena" (p. 110). As early as 1772 or 1773, in the introduction to the Physische Geographie, Kant had drawn the distinction between a logical division and a physical division of nature, had criticized Linnaeus' "system of nature, " and at least had touched on some of the problems involved in creating an "empirical system of nature. " In fact, as early as 1765 Kant had presented geography as an "empirical system of nature, " the purpose of which was to create a degree of unity at the empirical level, without which "all knowledge is merely piece-meal. " In the First Introduction to the Critique of Judgment, Kant is simply returning to an old problem.

It has been made clear earlier that geography is an "empirical system of nature, " and is the "empirical doctrine of nature. " It is geography that attempts to give us an integrated view of the world at the empirical level. In "empirical" physics, however, we likewise cannot deduce the "special" laws of nature from a priori principles; we must discover them in nature, although they are one and all subject to the categories of the understanding. Similarly, as we have seen, even the empirical laws of geography, in order to qualify as laws, must be subject to these categories. Nevertheless, the laws of "empirical" physics concern themselves with general and abstract characteristics of phenomena. The passages quoted above, however, indicate a concern with nature in the bedrock sense, as we experience it directly and in detail, in all its heterogeneity and diversity. It is geography that studies nature in its heterogeneity and diversity, in its spatial differentiation. Yet, the "empirical system of nature, " as a system, is not concerned with the unique, with the totally heterogeneous or with what appears isolated to us; it is concerned with the systematic, empirical arrangement of classes of phenomena. Classes, species, genera, etc. cannot be deduced from the categories of the understanding; and hence, another means is necessary to build up an "empirical system of nature" out of the empirically given. Reflective judgment does this. Through comparison of, and reflection upon, the heterogeneous empirically given, we can discern that some things in nature are, in certain respects, similar to some other things, and hence can be classified together. A true "empirical system of nature" cannot, however, proceed in the manner of Linnaeus, through the arbitrary logical division of a "whole," but rather it must proceed from the empirically given, guided by judgment.

There are evidently several important differences between physics as a "system of nature," and the "empirical system of nature." Physics deals with general and abstract characteristics of phenomena and exhibits these in their necessary conformity to universal a priori laws of reason; in no sense is it concerned with the heterogeneity and diversity of nature. Physics is concerned solely with similarities, and not with differences. On the other hand, the "empirical system of nature" begins with nature as differentiated, and attempts to discover similarities. But we must assume that nature in its differentiation does not differentiate itself in a totally unique fashion. If this were the case, it would be impossible to discover similarities among phenomena at the empirical level. Similarities must, so to speak, be embedded in the differentiations of nature. But this proposition is not empirically given, nor is it a lawful command of the understanding. Judgment provides the a priori principle that compels us to seek for an "empirical system of nature." However, since judgments are always "subjective," the classifications arrived at on the basis of reflective judgment, which is applied specifically at the empirical level, are therefore always empirically based and to some extent contingent. Physics then is a nomothetic discipline, whereas the "empirical system of nature," because of its necessarily subjective elements, is more akin to "art," something "where we realize a preconceived concept of an object which we set before ourselves as an end."[15] Hence, the "empirical system of nature" exhibits an idiographic quality, although it is certainly not idiographic in the sense of the "unique"; as pointed out above, as a system it cannot be concerned with the unique.

[15]Kant, Critique of Judgement, "Introduction," p. 34.

Or, to put it somewhat differently, physics is a "theory of nature," whereas any system constituted according to teleological concepts "is strictly speaking only incident to a description of nature."[16] Geography of course is a "description of nature." One of the crucial differences between a "theory of nature" and a "description of nature" is that the former can, and indeed must, present its propositions in mathematical form, and moreover, in the form of a logically integrated or deductive mathematical system. In addition, as we have already seen, mathematics is not applied to "a pure philosophy of nature in general," i.e., to a metaphysics of nature, but rather to "a pure doctrine of nature respecting determinate natural things," i.e., to a theory of nature. And since mathematical concepts are always constructed in intuition, they must apply to "determinate natural things" as given in experience; "apart from this relation they have no objective validity, and are a mere play of the imagination."[17] Nevertheless, despite the fact that mathematical concepts are constructed in intuition, they are one and all a priori concepts and hence can apply only to what comes under the jurisdiction of the understanding. Kant, however, has made it quite clear that at the strictly empirical level, the understanding can in no sense legislate "the possibility of a systematic experience," i.e., the systematic arrangement of all the particular empirical laws of nature. Hence, the "empirical system of nature," or any "description of nature" for that matter, cannot be represented in the form of a mathematical system, although a "description of nature" may of course use individual mathematical propositions incidentally, as aids to precision, as Kant does in the Physische Geographie.

In the last few pages an attempt has been made to build a strong case to support the contention that Kant has in mind, if not geography, at least something very like it, in his discussion of an "empirical system of nature." Has the case been made, and has the possibility been disposed of that Kant has "empirical" physics in mind, as Cassirer seems to believe,[18] or something else again? Probably not. Previous quotations respecting "genera," "species," and "classes" indicate that

[16]Ibid., Part II, sec. 79, p. 76.
[17]Critique of Pure Reason, A239 = B298.
[18]H. W. Cassirer, A Commentary, p. 209.

Kant probably had Linnaeus' "system of nature" in mind, [19] but because of his dissatisfaction with that system, he wanted one that was empirically grounded, i.e., akin to geography.

With respect to "empirical" physics, one more point is worth making. The "empirical system of nature" aims at creating an empirically based hierarchy of nature, consisting of orders, genera, species, subspecies, etc. On the other hand, Kant talks about "empirical" physics in terms of general concepts like "body." Although this concept has often been referred to, in previous contexts, as "universal," "general," or "abstract," it is nevertheless a rather "concrete" and empirical concept, since Kant applies it only to individual things. It represents one of the most general concepts we can employ to designate what, for purposes of physics, is common to an endless variety of particular things. It can thus stand for these particular things. This general applicability is what Kant seems to have in mind in his discussions of "empirical" physics. On the other hand, we cannot apply the concept "body" to "orders," "genera," and "species." These concepts do not designate or stand for particular entities. We cannot experience "orders," "genera," and "species," and since we cannot experience them, in the sense in which we can experience bodies, then laws, in the "objective" sense of that term, are not applicable to them. Hence, any "system of nature" of this sort is bound to remain a "description of nature." What these concepts represent for Kant is an attempt to organize the various forms and laws of nature, some more particular, some more general, into an hierarchical system. "Order," "genus," "species," etc., then, are "logical" concepts, that is, they are not concepts that are directly applicable to nature, but are needed for our own orderly comprehension of nature. Obviously, then, something much different is being done in constructing an "empirical system of nature" than is done in physics.

However, insofar as geography is to be understood as a lower level "system of nature" (i.e., one that studies some of the same objects as physics although at a more particular level), evidently something different is also being done in geography than in the "empirical system of nature" advocated in

<hr>

[19]Kant's reference to Linnaeus in the First Introduction to the Critique of Judgment, p. 20n, strengthens this view.

the First Introduction to the Critique of Judgment. Geography, as an empirical science of nature, although concerned with achieving an integrated view of the world at the empirical level, is concerned, like physics, with objects of experience, and not with "logical" concepts whose sole purpose is to aid our comprehension of nature at the empirical level. Kant shows no indication of abandoning his distinction between a logical division and a physical division of nature. Perhaps what he calls an "empirical system of nature" in the First Introduction to the Critique of Judgment could more reasonably be called a "logical system of nature,"[20] and hence the former title could be reserved for geography, since Kant has amply demonstrated on a number of occasions that geography is, strictly speaking, an "empirical system of nature."

Yet, the tasks of each of the three—"empirical" physics, the "logical system of nature," and the "empirical system of nature" or geography—remain somewhat different in their approaches to nature at the empirical level. The "special" laws of physics cannot be deduced from a priori principles; they must be discovered in nature. Such laws are concerned with general and abstract characteristics of phenomena, and with the similarities of nature. However, these laws, although all subject to the a priori laws of the understanding, apparently cannot be exhibited by that faculty in their systematic relations to one another. Geographical laws are more specific, particular, and contingent than the laws of "empirical" physics, and are concerned with the diversity of nature, with nature as we experience it in its spatial differentiation. In addition, however, unlike "empirical" physics, geography attempts to give us a systematic view of the world at the strictly empirical level. How can geography achieve this? In addition to requiring the idea of a "whole" which sets the limits for the discipline, and certain postulates of empirical thought, geography apparently also requires principles of judgment. Principles of judgment, since they are an a priori requirement, are therefore a necessary presupposition for the possibility of systematic scientific inquiry at the empirical level. The "logical system of nature" has two aims: (1) it aims at a classification of the

[20]In fact, Kant occasionally employs the designating "logical system of nature" in place of "empirical system of nature." See, for instance, First Introduction to the Critique of Judgment, p. 20.

various objects of nature; and (2) it aims at an hierarchical arrangement of all the empirical laws of nature, at the subsumption of the most particular laws under more general ones, and so on. Kant gives no indication that it was ever the aim of geography to achieve an hierarchical order of nature. Yet, it certainly employs "logical" concepts such as "class," "species," "physics," "commerce," "theology," etc. The possibility of achieving a classification of the objects of nature, as prescribed by judgment, would thus appear to be a necessary prerequisite or propaedeutic for geography as well as for "empirical" physics. On the other hand, the achievement of a systematic hierarchy or order of nature would appear to be more an end-product of knowledge, since the laws of nature would need to have been discovered before they could be so ordered. Such a "system of nature" would presumably include the laws of both geography and "empirical" physics.

THE PLACE OF GEOGRAPHY IN A CLASSIFICATION OF THE SCIENCES

Kant's classification of the sciences can best be introduced by means of a simplified illustration (see Fig. 1). The diagram, however, requires several qualifications. It does not include everything that Kant regards as constituting knowledge,[21] but only what comprises science in the strict or narrow sense. Hence, it omits metaphysics, i.e., any knowledge devoid of empirical content, pure mathematics, and ethics.[22] In addition, the diagram leaves out "common knowledge," i.e., purely empirical knowledge that cannot constitute science for Kant. Furthermore, the diagram may be somewhat misleading since it sets the empirical sciences directly under the theoretical sciences, and thus could convey the idea of a hierarchy. Actually, there are four distinct levels in Kant's classification, two of which do not appear in the diagram since they do not constitute

---

[21]For a diagram that illustrates the whole range of knowledge for Kant, see Norman Kemp Smith, A Commentary to Kant's "Critique of Pure Reason," p. 580.

[22]In order to maintain Kant's parallelism of the sciences, I have included in my diagram psychology with a question mark, although, as indicated earlier, it is questionable that Kant leaves any room for a psychology which could be considered a theoretical discipline possessing an empirical component.

	NATURE (Outer and inner sense)	MAN (SOUL or SELF) (Inner sense)
THEORETICAL	PHYSICS and OTHER THEORETICAL NATURAL SCIENCES	PSYCHOLOGY ?
EMPIRICAL	GEOGRAPHY and HISTORY OF NATURE (SPACE and TIME)	ANTHROPOLOGY and HISTORY OF MORALS (TIME)

Fig. 1. Kant's Classification of Sciences

science in the strict sense. The highest level is that of meta-physics in general, either of nature in general or of self in general, wherein the synthetic a priori principles of ontology and ethics respectively are studied. The second level com-prises the theoretical sciences, which aim at the establish-ment of "universal" abstract statements of law respecting nature and self that possess empirical content. The third level, that of the empirical sciences, aims at the establishment of partic-ular concrete empirical laws respecting nature and self. The fourth level, that of "common knowledge," is purely empirical and fact-finding.

Perhaps the most interesting issue raised by Kant's clas-sification of the sciences is his idea of "levels." The levels are not entirely discrete, since some overlap or contact exists be-tween them. Nevertheless, Kant intends them as distinct levels in the sense that they do not form a hierarchy, i.e., one cannot pass downwards from level to level deductively, nor upwards from level to level inductively. For instance, the general laws of physics must be discovered in nature, and are not deducible from universal synthetic a priori principles of the understanding, although they must all be exhibited in conformity to those prin-ciples; nor can one arrive inductively at synthetic a priori prin-ciples of understanding on the basis of a compilation of general laws in physics. Similarly, the more specific laws of geography must also be presented in conformity to universal synthetic a priori principles of understanding, although they are not deducible from them, nor, of course, could such principles be derived inductively from the specific statements of geography. While contact is thus maintained between the first and third levels, it is more difficult to exhibit any essential distinction between the second and third levels. Since scientific statements at both these levels must conform to synthetic a priori principles of understanding, obviously no differentiation is possible on this basis. Nevertheless, although geography serves as a propaedeutic to physics, in that it studies some of the same objects as does physics, Kant would probably maintain that we cannot arrive at the "universal" abstract statements of physics on the basis of an induction from the much more limited and regional statements of geography. Since the statements of geography, by their very nature, are regional and thereby individually restricted in extent of empirical coverage, the compilation of geographical statements can never yield more than general statements of limited regional truth. However, are not the more specific or restricted statements of geography deducible from the more general statements of physics? Can they not be exhibited as instances of such statements? This issue may present the greatest difficulties to maintaining Kant's idea of "levels."

Undoubtedly, the particular and concrete terms of geography could be translated into the general and abstract terms of physics, such as "body," "state," "motion," etc., and thus the statements of geography could be presented as particular instances of more general physical statements. Nevertheless, although there seems to be no textual evidence in support of the point, I think Kant would have to maintain that deduction in this context is an after-the-fact accomplishment. That is, in the actual course of scientific investigation we cannot simply deduce the specific and restricted regional statements of geography from the "universal," abstract statements of physics, but rather, further empirical research is necessary to discover geographical generalizations. Concepts such as "body," "state," and "motion" are not a priori concepts, however, but abstractions from reality, and in this sense, a link is certainly maintained between the second and third levels. In addition, Kant

149

undoubtedly believed in the possibility of constructing a hierarchy of the forms and laws of nature. Yet the hierarchy is the accomplishment of judgment, not of understanding; hence, it is not "objective," but rather a "subjective" requirement of human reason. With respect to the fourth level—"common knowledge"—facts are obviously required at the second and third levels, and thus in a sense the fourth level can be thought of as linked to the second and third, although "common knowledge," strictly speaking, is concerned only with cognitions in their isolation from one another. However, particular facts can never be deduced from general scientific statements, nor can the mere compilation of facts through induction ever lead, of itself, to the establishment of scientific statements. Hence, the fourth level remains distinct.

SUMMARY OF KANT'S CONCEPT OF GEOGRAPHY

Geography as a science is confined to a study of natural phenomena occurring on or near the surface of the earth. As distinct from physics, it is an empirical science, not a theoretical one. But since geography provides systematic knowledge of nature, it is also a "system of nature," and a law-finding discipline. Although it studies some of the same objects as does physics, geography is a lower level "system of nature," which implies that it studies the relations among particular and concrete things rather than among abstract and general characteristics of things, and that it concentrates upon the differentiations rather than upon the similarities of nature. Hence, geographical propositions cannot be expressed, as can those of physics, in the form of a thoroughgoing deductive mathematical system. Geographical knowledge then is to some extent "rough," and is acquired either through direct observation or through the employment of hypotheses and analogies, procedures which are peculiar to empirical sciences.

At the empirical level, geography as a science is contrasted with anthropology and history. Geography is based on "outer sense" and concerns itself with nature, whereas anthropology is based on "inner sense" and concerns itself with man in the sense of soul or self. Together, geography and anthropology form a "whole" of our pragmatic, empirical knowledge of the world. But geography is also concerned with man insofar as he can be considered a part and product of nature. Geography deals with the external physical characteristics of man that do not ordinarily condition human self-consciousness, and with customs in the sense of unreflectively held mores. The distinction between geography and history is a secondary one that pertains basically to "outer sense"; hence, in relation to geography, history is entirely the history of nature. Geography is concerned with relations among things that coexist in space, whereas history is concerned with the sequences of natural events in time. The distinction between the two, however, is not absolute. Since history always presupposes geography, the latter is ontologically prior to the former, whereas, since geography requires historical knowledge for explanatory purposes, it is epistemologically dependent on history.

Pragmatic and teleological aspects of geography can also be distinguished. As an academic discipline, geography functions pragmatically in two respects: (1) as a propaedeutic to both science and life; and (2) as an end-product of knowledge. As a propaedeutic to science, geography provides an elementary introduction to physics and other theoretical natural sciences; as a propaedeutic to life, it provides knowledge of the surface of the earth, the stage upon which man acts out his life. As an end-product of knowledge, geography is synoptic, and attempts to provide us with a simplified yet integrated and meaningful view of the world as a whole, which no specialized discipline can do. Teleological principles play a role in geography in two respects. (1) They function as heuristic devices, since the attempt to view nature as designed may lead us to make a number of discoveries respecting the relations among things in the space of the earth's surface. (2) As reflective, they enable us to build up an integrated view of the world at the empirical level, since only through reflection upon the heterogeneity and diversity of nature as given in experience can we discern that some things are in certain respects similar to some other things in nature, and can be classified together. Only in this way can we eventually create a system out of the diverse forms and empirical laws of nature.

VII
The limits and scope of contemporary geography

The concluding chapters of this study can in no sense be regarded as constituting an adequate or exhaustive discussion of
the nature of geography, although the question of whether or not
geography exists as a distinct scientific discipline will be explored. Moreover, neither a history of contemporary geographical thought, nor an empirical inventory of types of geographical
literature is intended. The purposes of this study will be served
if the positions discussed have been held by only one person.
The discussion will be limited primarily to topics that relate
closely to Kant's concept of geography. Nevertheless, given
even these limitations, at least three major topics require discussion: (1) the limits and scope of contemporary geography;
(2) contemporary concepts of the nature of geography and, more
generally—what kind of science geography is; and (3) the place
of geography in a classification of the sciences. Something
needs to be said also concerning the concept "region, " the relations between geography and history, and explanation in geography.

THE LIMITS AND SCOPE OF
CONTEMPORARY GEOGRAPHY

In attempting to determine the extent to which Kant's concept of
geography can be regarded as an adequate foundation for contemporary geography, one is faced with an initial major difficulty. Because of developments in science and the profuse
growth of various new sciences since Kant's day, his concept
of the limits and scope of geography is inevitably much broader
than any contemporary concept can reasonably be. Hence the
scope of geography as conceived by Kant embraced much material which we would today regard as the prerogative of disciplines other than geography. For instance, Kant's geography
included oceanography, which today is almost certainly regarded
as a distinct discipline, or group of disciplines, concerned with
the study of various physical and chemical properties of ocean
water, bathyography, marine life, and even economic aspects
of marine life.[1] What Kant called the study of "ancient" states
of the earth, which he regarded as part of geography, is today
the concern of historical geology. In addition, much of his physical geography in the narrow sense, i.e., the physical surface
of the earth, would today be included under geology. Much of
what Kant included in the study of climate is today the concern
of the meteorologist. The status of biogeography, which he included as an important branch of physical geography, is at
least highly questionable today as a recognizable branch of
geography; it might appear to be much more reasonably the concern of biologists. Kant included the study of race under physical geography; today that theory is the concern of the biologist
and the anthropologist. What he called "moral geography," the
study of customs and mores, is today divided between anthropology and sociology. His "theological geography" we would today
incline to call comparative sociology of religion. The "mathematical geography" of Kant's day is now the concern of either
astronomy or of that highly specialized branch of mathematics
known as geodesy. In fact, aside from regional geography, the
only one of the special geographies listed by Kant that bears a

---

[1]Oceanography is perhaps better thought of as a synoptic discipline, since few if any
scientists are trained in oceanography as such. Rather, scientists come to the study
of the ocean, a major and distinct portion of the earth's surface, from a background
of training in one or more of physics, chemistry, biology, geology, etc.

close relation to contemporary geographical pursuits is "commercial geography," but even in this case, the study of international trade—one of the important aspects of his "commercial geography"—is certainly a major branch of economics. In short, one may legitimately ask whether or not the science of geography, as understood by Kant and his contemporaries, any longer exists, and if it has indeed not been carved up and divided among a number of more specialized sciences that have arisen and established their domains since Kant's day. But, if this is the case, one may reasonably ask if a new scope, one not envisaged by Kant, has arisen for geography. Or has geography really ceased to exist as a separate discipline? Does it, in fact, consist of anything more than bits and pieces left over from other sciences—bits and pieces that are difficult to fit into the bodies of knowledge that comprise a variety of specialized, contemporary sciences? It is nevertheless hard to imagine that a discipline which has been recognized for over two thousand years is about to disappear altogether.

There appears to be almost universal agreement between Kant and contemporary geographers that geography is confined to a study of the surface of the earth, although the precise limitations of the concept "surface of the earth" are far from certain.[2] The basic problem, however, of determining the limits and scope of geography from a contemporary point of view is that any number of sciences other than geography also study phenomena that are limited to the surface of the earth. Leaving aside "universal" sciences such as physics, and highly experimental sciences such as chemistry, the sciences which are primarily confined to studying phenomena occurring on or near the surface of the earth include geology, oceanography, biology, history, and the various social sciences. How then are we to distinguish geography, as a science of the earth's surface, from various other sciences that study phenomena confined to the surface of the earth? If we look at the issue in terms of the traditional distinction between physical, biological, and human, it would appear that these three major divisions of knowledge, as regards the surface of the earth, have been pre-empted by geology, biology, and the various social

[2]For recent discussion of this point see Hartshorne, Perspective on the Nature of Geography, chap. iii, "What Is Meant by the 'Earth Surface'?" pp. 22-25.

154

sciences. If this is the case, what is there left for geography to do?

It would be useful to examine somewhat more closely a number of contemporary sciences dealing with the surface of the earth and their relations with geography. As remarked earlier, it would appear that geography is not concerned with the whole of the surface of the earth, since oceanography has claimed as its special prerogative the study of the oceans, which occupy approximately 70 per cent of the earth's surface. In this respect, contemporary geography is much closer to Francis Bacon's concept of geography than to Kant's, since, for Bacon, the "ge" in geography means literally "earth" or "land," and therefore excludes the ocean.

THE PHYSICAL AND BIOLOGICAL SCIENCES

Is there any meaningful basis upon which to draw a distinction between geology and geomorphology conceived as a branch of geography? The study of the composition and stratigraphy of the "hard" or "solid" rock surface of the earth is clearly the concern of geology. But the concept "rock" is often defined more broadly to include the waste mantle and glacial drift of the earth's surface, exclusive of soil, an organic and inorganic mixture, the detailed study of which in its present state is the concern of neither geology nor geography, but of pedology. When geology concerns itself with the study of the Pleistocene epoch, it is studying primarily glacial landforms composed of till and other forms of glacial debris, and not "solid" rocks; and the geologist who specializes in the most recent geological epoch is confined largely to the study of sedimentary deposits. In these areas, considerable duplication of work exists between geologists and geomorphologists, and it is difficult to discover any rational ground for distinguishing the two. At least three positions are held in current debate over the relations between geology and geomorphology. Geologists who are concerned with geomorphology tend to regard it as a branch of geology. The following is a fairly typical remark:

Geomorphology is primarily geology, despite the fact that some geomorphology is taught both in Europe and in this country as a part of physical geography. In most geography courses landforms are treated rather incidentally as a part of the discussion of the physical environment of man, but emphasis usually is placed upon man's adjustments to and uses of land forms rather than upon land forms per se.[3]

Here, the distinction between geology and geography is made on the grounds that the latter is concerned with the study of relations between man and his physical environment. Hence, since geomorphology is concerned with the study of landforms per se, it is ipso facto a branch of geology, the science which studies the physical surface of the earth.

On the other hand, geographers who are concerned with geomorphology tend to regard it either as a legitimate branch of geography or as a science that has emerged in its own right, independent of both geology and geography. As an example of the first case, B. W. Sparks[4] bases his position initially on W. M. Davis' argument that the evolution of rock structure belongs to geology whereas the wearing down of rock structure belongs to geography. Even so, according to Sparks, geomorphology, understood as the study of landforms that we can actually visualize as present on the earth's surface here and now, has three recognizable areas of study: (1) the study of the relations between landforms and underlying rocks; (2) the study of the evolution of landscapes that are present on the earth's surface; and (3) the study of the actual processes of erosion that give rise to landforms. Although (1) tends to be neglected by both geologists and geomorphologists, it would appear to lie more clearly within the province of the former, since the basis of study is the underlying rock structure. (2) involves a considerable overlap with stratigraphy, and hence can be regarded most logically as the tail-end of historical geology. (3) is most clearly geomorphology as distinct from geology, and in fact the study of fluvial processes in erosion, for instance, or of slopes, is generally much more the concern of people who call themselves geomorphologists than of geologists. Nevertheless, the attempt to draw a sharp distinction between depositional and erosional processes can be fraught with logical difficulties. This is especially the case with the study of glacial landforms,

[3]William D. Thornbury, Principles of Geomorphology, p. 1.

[4]Geomorphology, chap. i, "The Aims and Position of Geomorphology," pp. 1-6.

many of which are the end result of both types of processes. To divide the study of a particular landform between two distinct disciplines strikes one as a highly arbitrary move that could easily interfere with scientific comprehension. Although Sparks berates as "pointless" the debate over whether geomorphology is geology, geography, or an independent discipline, he evidently sides with the "wearing down" half of Davis' distinction, and hence regards geomorphology as a legitimate branch of geography.

The following remark may be taken as fairly typical of one who regards geomorphology as a distinct science in its own right:

> . . . the emphasis of geomorphology proper differs from the emphasis of geology or of geography. Most geologists are concerned with the formation and deformation of rocks, geomorphologists with the erosional facets by which rocks are truncated. Geographers as a group deal with man in relation to his environment. As specialists, geomorphologists examine certain aspects of that environment, without necessarily considering man at all.[5]

Others, however, would regard this debate among the various physical sciences of the earth, respecting their private domains and even their claims to independence, as a mere tempest in a teacup. The recent move in several universities to reinstitute a general physical science of the earth is informative in this respect:

> It became obvious to us that the study of the earth and its environment is best handled by all concerned with it, and that rigid, old-fashioned subdivisions served no purpose other than to stifle knowledge and obstruct free interchange of ideas and information. Much more than an administrative reshuffling of faculty was involved. The change created a more logical and natural grouping of current scientific faculty and, at the same time, a structure perhaps better attuned to the acceptance of new people who may transgress the narrower or more classically defined boundaries of the older departments.[6]

The new geophysics includes, among other specialties, geophysics in the old sense, geology, geomorphology insofar as it is a generalizing quantitative science, and meteorology. Gone is the old division of the earth into interior, surface, and atmosphere. The physical earth, including its atmospheric

[5]G. H. Dury, The Face of the Earth, p. 2.

[6]Julian R. Goldsmith, "The New Department of the Geophysical Sciences," Chicago Today, I, No. 2 (1964), 4-14; quote from p. 5.

envelope, is to be regarded as a whole, to be studied by a single discipline, although specialities or divisions within the single discipline will still be recognized. Geography, insofar as that discipline is a social science, or studies the relations between man and his physical environment, is excluded. The new geophysics can be thought of as paralleling biology, in the sense that although we recognize that biology can be sub-divided into several dozen clearly recognizable specialties or areas of scientific concentration we nevertheless regard it as a single science or discipline with a fairly clearly demarcated area of study.

Are there any grounds for distinguishing climatology from meteorology? The former is generally regarded as a branch of geography, the latter a branch of physics. Meteorology, understood as the physics of the air, is a specialized branch of physics that takes for its province the study of the earth's atmosphere in all its physical aspects. Climatology, on the other hand, is not concerned with the atmosphere as such, but with the surface of the earth, or more accurately with inter-action between the earth's surface and the atmosphere which gives rise to weather and climate. In principle there is evi-dently a focal difference between the two, in that the climatol-ogist has little interest in, and little competence for, the detailed analysis of the physics of the atmosphere, being pre-pared to take on faith the findings of the meteorologist, whereas the meteorologist ordinarily has little knowledge of the features of the earth's surface. In actual practice the logician is faced with a major difficulty in attempting to distinguish the two ra-tionally. This difficulty arises from the fact that the study of weather, the day to day changes in the interaction between the atmosphere and the earth's surface, has traditionally been the concern of the meteorologist, and one of his major roles is that of weather-forecaster. We cannot, however, in any strict sense, deduce the actual weather of a particular place on the earth's surface from the principles and laws of atmospheric physics, nor could we do so even if those laws and principles were far more accurately understood than they are today. The variability of the features of the earth's surface, both physical and artificial, creates an almost endless variety of local con-ditions, and hence it can be argued that the accuracy of local weather forecasts would be improved considerably if the meteorologist possessed a greater knowledge of the various features of the earth's surface. This issue, however, is complicated by the recognition of subdivisions of the major disciplines such as micrometeorology and microclimatology. Since micrometeorology concerns itself with the principles and laws of the atmosphere very close to the earth's surface, principally as the atmosphere has been substantially influenced by the surface of the earth, one may presuppose of the micrometeorologist a detailed knowledge of that surface. On the other hand, the study of climate, the average yearly patterns of weather, has generally been more the concern of the climatologist than of the meteorologist; yet here, we are better able to deduce broad climatic patterns than particular weather conditions from our knowledge of astronomical and atmospheric principles. This is not, of course, entirely the case. If the surface of the earth were completely uniform, then there could be no science of climatology, since climates, aside from initial questions of classification, could be distinguished entirely on deductive grounds alone. Insofar as the climatologist is concerned with spatial patterns of climate on the earth's surface, and their implications for plant and animal life and human activity, he is clearly doing something geographical. Although there is prima facie evidence to suggest that the climatologist, in the main, is concerned with matters that differ from those that concern the meteorologist, the logical basis for clearly distinguishing the two as separate disciplines is far from evident and in fact is even confused.

Is there any basis for distinguishing biogeography as a branch of geography from biogeography as a branch of biology? Studies of spatial and environmental aspects of plant and animal life would appear to be well entrenched as elements of biology. For instance, spatial studies that are almost the exclusive concern of the biologist include the territorial rights of mammals, birds, and fish, and the migration of birds and mammals. In addition, varied types of biological studies are concerned with plants and animals in their relations to any number of environmental factors. The branch of biology known as "phenology" specializes in the study of the relation of plants and animals to weather changes, including seasonal changes.

Biologists also study the adaptations of insect life to changes brought about by man's artificial environment.[7] Biologists have long been concerned with questions of conservation, the preservation of natural habitats, and the effects of the invasion of natural habitats by man and other alien species.[8] When one gets into the area of studies that are usually labelled "ecology," a wide range of non-biological factors are introduced, yet these studies are often best handled by biologists, since their central focus remains the biological species and subspecies, and thus presupposes a thorough knowledge of them. One biologist has recently remarked: "This is a field which calls for knowledge of a wide range of subjects. The ecologist must be a map reader, something of a geologist and a student of soils and climate."[9]

By the term "biogeography," however, geographers often understand the study of the large-scale spatial patterns of plant and animal life as these occur on the earth's surface. Yet, even in this area, much of the research is carried on by biologists, primarily because the central focus of study is biological, and thus presupposes years of study required to obtain a knowledge of a variety of biological species. Nevertheless, the explanatory basis of large-scale biogeographical patterns lies outside biology, principally within climatology. This is especially the case with plants; as one biologist who specializes in plant geography has put it, "climate is the master."[10] While this might suggest that the understanding of large-scale biogeographical distributions falls most readily within the competence of the climatologist, a knowledge of current climatology can only supply reasons for the ability of plants and animals to maintain their habitats under present climatic conditions. An adequate explanation of present bioge-

[7]For instance, see H. B. D. Kettlewell, "Industrial Melanism," Animals, V, No. 20 (1965), 540-543.

[8]As an example, see Julian Huxley, The Conservation of Wild Life and Natural Habitats in Central and East Africa.

[9]Ben Dawes, A Hundred Years of Biology, p. 342. A recent example that illustrates the complexity of such studies, even on the micro-level, is J. R. Lewis, The Ecology of Rocky Shores. In this work, the author studies the interaction of tide, temperature, salinity, desiccation, and exposure as they influence the production of a variety of fauna and flora, and control their locations; the work also studies the colonization and distribution of specimen sites.

[10]Nicholas Polunin, Introduction to Plant Geography, p. 9.

ographical distributions over the surface of the earth can only be reconstructed historically from the evidence supplied by palaeontology and historical geology.[11] In addition, the whole question of "geographical speciation" is apparently so vital an aspect of the theory of evolution[12] that its removal from biology and incorporation in geography would be as undesirable as removing the theory of evolution itself from biology and locating it within history on the grounds that its perspective is primarily historical. In this connection, Simpson[13] has stated:

> It is a proper and, indeed, indispensable aim of biology to explain evolution in theoretical terms. Such explanation, in this as in any field of science, necessarily involves abstraction, with operations at one, two, or still more removes from the objective data. In considering theories of evolution, it nevertheless must be kept in mind that such theories, to have possible validity, must arise from and then in turn apply to real events, involving individual living things, occurring in a unique temporal sequence, and played out on and near the surface of a highly concrete body—our planet Earth.

Occasionally, however, geographers have argued that biological classifications of animal life fail to serve geographical purposes. One geographer has recently gone so far as to declare that: "We have a strong impression that zoogeography takes little account of geographical problems."[14] The author goes on to argue that animals can be divided into ocean, lake, and river species, and into forest, grassland, and other vegetative environmental species, and furthermore into subdivisions of these general categories. However, just how this kind of classification enlarges our biogeographical knowledge is not made clear. Carl Sauer[15] (following Hettner) suggests that the geographer may regard vegetation and animals as part of the human habitat, and hence consider them in terms of, for instance, economic plant and animal geography. In this case, though, the central concern is no longer biological or biogeographical but economic or economic-geographical.

[11]See Charles Singer, A History of Biology, pp. 272-273; and T. A. Goudge, The Ascent of Life, pp. 67-68.

[12]A. J. Cain, Animal Species and Their Evolution, especially chap. viii, "Geographical Speciation."

[13]George Gaylord Simpson, The Geography of Evolution, preface.

[14]De Jong, Chorological Differentiation, p. 152.

[15]"The Morphology of Landscape," in John Leighly, ed., Land and Life: A Selection from the Writings of Carl Ortwin Sauer, p. 341.

THE SOCIAL SCIENCES

When one turns from the physical and biological sciences to the social or human sciences, and seeks to secure a place for geography in relation to them, one is perhaps faced with even greater difficulties. This is primarily due to the fact that the social sciences are much less developed, and hence greater uncertainty exists as to the precise boundaries among them, and the scope and limits of each. Ernest Nagel[16] has recently argued that much of social science literature consists either of social theories that are really a species of social philosophy, i.e., speculation about the social order or social system that lacks empirical warranty, or of a variety of particular empirical studies that lack theoretical justification. In short, much of social science theory lacks any "fit" with the world of reality or the world of experience. But surely one of the principal means of securing a "fit" is to pay much greater attention to the general spatial and temporal dimensions of the economy or the society. Such an undertaking inevitably involves the social scientist with geographical and historical matters. In considering the geographical involvement of social scientists, we will look briefly only at the relations between economics and economic geography, and between urban sociology and urban geography.

Alfred Weber, one of the early leaders in the study of the general theory of the location of economic activity, would certainly support Nagel's contentions respecting the problem of securing a "fit" for the social sciences. Weber argues that

the question of the location of industries is a part of the general problem of the local distribution of economic activities. In each economic organization and in each stage of technical and economic evolution there must be a "somewhere" as well as a "somehow" of production, distribution, and consumption. It may be supposed that rules exist for the one as well as for the other. Still, political economy, in so far as it goes beyond the analysis of elemental facts and beyond pure theory, is of necessity primarily description and theory of the nature of economic organization. The presentation and theoretical analysis of the nature, the sequence, and the juxtaposition of the different kinds of economic organization is its natural content as soon as it attacks concrete reality. This is such an enormous content that it should not occasion wonder if a young science, limiting itself in its initial tasks, treats the "somewhere" of economic processes simply as a function

[16]The Structure of Science, pp. 447-450.

162

of the nature of that process, although the location of an economic process is only partly a function of its nature. In other words, political economists have dismissed this problem of location with some general references to rules of local and international division of labor, etc., or political economy has left to economic geography the theoretical consideration of the distribution of economic processes over a given area.[17]

In brief, Weber is pointing out that once the economist goes beyond pure theory and the analysis of elementary economic facts, he inevitably and necessarily gets involved with questions concerning the location, spatial distribution, and temporal development of a variety of economic activities, or, he becomes involved with geographical and historical matters. In other words, in order to secure a "fit" or a justification for economic theory, the economist must inevitably involve himself with the geographical and historical dimensions of economic matters. In addition, Weber is arguing that this involvement with the geographical and historical dimensions signifies a growing maturity for economics as a science.

Weber goes on to argue that economic geography, by its very nature, cannot achieve the desired result of formulating a general theory of economic location:

Naturally, the latter [economic geography] is able to approach the problem only in so far as it can be explained by purely physical facts. The result is as unsatisfactory as if we had left the analysis of the nature of economic processes, i.e., political economy, to the technical sciences [p. 1].

Weber, in this context, does not explain what he means when he limits economic geography to a consideration of "purely physical facts." However, in a later context (pp. 20-25), he sheds some light on the issue by remarking that "locational factors" are to be understood in terms of "advantages of cost," which is a purely economic matter (p. 25). Since the concern of a general theory of the location of economic activity is confined to "purely" economic matters, then "industry is agglomerated or spread within its geographical network according to certain general rules which are quite independent of geography" (p. 21). In addition, Weber points out that "there is no method by which one could deduce from known premises the special locational factors which exist for given industries on account of natural or technical peculiarities" (p. 23). Although Weber is never

[17]Alfred Weber, Theory of the Location of Industries, p. 1.

really clear about the matter, it appears that the role of economic geography, at least on the question of industrial location, is confined to the study of the physical background of economic location, i.e., to topography and climate, especially as they influence transportation, and to the study of the special or individual factors that govern the specific locations of particular industries; whereas the general theory of economic location, since it is expressed in economic terms, is purely the concern of the economist.

Lösch went much further than Weber when he proposed "to view all economic activities geographically. In principle all economic theory can be reformulated thus from a spatial aspect."[18] Lösch was well aware of the fact that he was doing something geographical in the broad sense of that term, yet he regarded his undertaking as falling entirely within the scope and competence of economics. Since, in principle, all economic theory can be restated in spatial terms, to surrender the spatial dimension of economic theory to the geographer would amount to surrendering the whole of economics. Presumably, as in Weber's case, the function of economic geography would be confined to studying the particular or individual or non-general aspects of the economy in space.

Nevertheless, despite Lösch's ambitious intentions, many theoretical aspects of economics continue to be studied on a largely non-spatial basis. These would include such matters as the general study of markets, costs, production, consumption, demand and supply, monopoly, competition, money, interest rates, investment, balance of payments, exchange rates, etc. The general study of these and certain related matters is what is usually regarded as the central concern of the economist. On the other hand, economic geographers have traditionally been concerned with studying much more down-to-earth matters, such as, for instance, the actual conditions under which, and the actual places in which, a variety of agricultural goods are produced, or the actual conditions of trade among various nations over the surface of the earth. Robert McNee, an economic geographer who is interested in the relations between economics and economic geography, has recently pointed out that the traditional basis for distinguishing the two has been

[18]August Lösch, The Economics of Location, p. xiii.

that economists displayed little concern with the actual spatial dimensions of economic activity, whereas economic geographers have been largely preoccupied with such matters.[19] However, the issue is not so simple. As McNee points out, many economists, because of the occurrence of major events such as the great depression and growing concern over the establishment of viable economies in a number of underdeveloped nations, have inevitably become more interested in the actual spatial dimensions of economic matters. In fact, some economists have argued that the economic generalizations built up on the basis of preoccupation with advanced capitalist economies have little specific application to the conditions faced by the underdeveloped economies of the world.[20] And, when one comes to consider the literature produced in some of the more specific branches of economics, such as agricultural economics, one is aware of a great similarity with the primary interests of some economic geographers.[21]

On the other hand, some geographers are becoming increasingly preoccupied with general economic location studies.[22] The situation is further complicated by the recent appearance of a "new" discipline known as "regional science," which concerns itself primarily with spatial and locational

[19]McNee, "The Changing Relationships of Economics and Economic Geography," Economic Geography, XXXV, No. 3 (1959), 189-198. However, McNee points out that Adam Smith, the acknowledged founder of the science of economics, is a notable exception to the generalization, since he often concerns himself with specific geographical matters.

[20]See, for instance, Albert O. Hirschman, The Strategy of Economic Development.

[21]Some recent examples of literature in agricultural economics that are virtually indistinguishable from literature in economic agricultural geography include: Colin Clark and Margaret Haswell, The Economics of Subsistence Agriculture; Ester Boserup, The Conditions of Agricultural Growth; and Thomas T. Poleman, The Papaloapan Project: Agricultural Development in the Mexican Tropics. Clark and Haswell (pp. 168-170) argue that the development of adequate transportation facilities is the most vital factor needed in securing a viable agricultural basis for the economies of underdeveloped nations, and they evidently regard the study of transportation facilities as of central economic importance. Yet, the study of transportation facilities has, in recent years, become a matter of increasing interest to a number of geographers.

[22]See, for instance, Report of the Ad Hoc Committee on Geography, The Science of Geography, "Location Theory Studies," pp. 44-49. In this work von Thünen, Alfred Weber, and Lösch, among others, are regarded either as forerunners of contemporary geography or as geographers, despite the fact that none of them so regarded himself but rather each looked upon his work as a legitimate economic pursuit.

factors of the economy.[23] Because of these developments in both economics and economic geography, the two appear to be drawing closer together. Yet this "drawing together" can mean, logically, only that economic geography is in the process of being absorbed by economics, since the theoretical basis of the discipline is to be found in economics, and since there are no logical grounds upon which to exclude the spatial aspects of the economy from economics. This is not to deny that greater concern with the geographical and historical dimensions of the economy may not bring about significant changes in economic theory.

With regard to the relations between urban sociology and urban geography, an interesting article has recently been written by Ernest W. Burgess, one of the founders of the Chicago school of urban sociology.[24] He divides the study of urban centres into two aspects:

First, their spatial pattern: the topography of the local community; the physical arrangements not only of the landscape but of the structures which man has constructed, that sheltered the inhabitants and provided places of work and of play.

Second, their cultural life: their modes of living, customs, and standards [p. 7].

Burgess goes on to point out that the first type of study gave rise to what became known as "ecological" studies, which included "all that could be mapped; the distribution, physical structures, institutions, groups, and individuals over an area" (ibid). In fact, much of the early work of the Chicago school of urban sociology was of this type, since it was necessary to possess an initial knowledge of the whole physical layout and spatial interaction of the city.

[23]As one observer has pointed out, most of the leaders of the "regional science" movement were originally trained as economists, and hence, through adaptation of traditional economic methods to the space-economy, "regional science" to date consists of little more than spatial or regional economics (Lloyd Rodwin, "Regional Science: Quo Vadis?" Papers and Proceedings of the Regional Science Association [hereinafter referred to as PPRSA], V [1959], 3-20). I can myself discover no grounds, either logical or real, for as yet regarding "regional science" as a distinct or separate science in its own right.

[24]Ernest W. Burgess and Donald J. Bogue, "Research in Urban Society: A Long View," in Contributions to Urban Sociology, ed. by Burgess and Bogue, pp. 1-14; see especially the section entitled "A Short History of Urban Research at the University of Chicago before 1946" (pp. 2-13), which was written by Burgess.

One historian of sociology has recently argued that the great merit of human or social ecology was that "of having drawn the attention of sociology to the fact that individuals and their social institutions do exist in a spatial environment, which, though largely the result of social factors, appears to them as natural, on which they are dependent, and which impresses itself on everything they think and do."[25]

If urban sociology is vitally and necessarily concerned with "spatial patterns" and the physical layout of cities, what is there left for urban geography to do? Robert Park, the acknowledged leader of the human ecology movement, draws a distinction between human ecology and geography in the following terms:

> Ecology, in so far as it seeks to describe the actual distribution of plants and animals over the earth's surface, is in some very real sense a geographical science. Human ecology, as the sociologists would like to use the term, is, however, not identical with geography, nor even with human geography. It is not man, but the community; not man's relation to the earth which he inhabits but his relations to other men, that concerns us most. . . .

> Geographers are probably not greatly interested in social morphology as such. On the other hand, sociologists are. Geographers, like historians, have been traditionally interested in the actual rather than the typical. Where are things actually located? What did actually happen? . . .

> Human ecology, as sociologists conceive it, seeks to emphasize not so much geography as space. In society we not only live together, but at the same time we live apart, and human relations can always be reckoned, with more or less accuracy, in terms of distance. In so far as social structure can be defined in terms of position, social changes may be described in terms of movement; and society exhibits, in one of its aspects, characters that can be measured and described in mathematical formulas.[26]

Geography, then, is not concerned with distance or spatial relations, but is to be distinguished from human ecology on the grounds that it studies (1) the relations between man and his physical environment and (2) the actual rather than the typical factors of location.

However, few urban and human geographers would be satisfied with this exclusion from considerations of distance and spatial relations. Burgess' distinction, quoted above, perhaps offers a better contemporary basis for drawing a boundary be-

[25]Heinz Maus, A Short History of Sociology, p. 127.

[26]Robert E. Park, "The Urban Community as a Spacial Pattern and a Moral Order," in The Urban Community, ed. by Burgess, pp. 3-4.

tween urban geography and urban sociology, if we allow that study of what he calls the "spatial pattern" is more geographical, whereas study of "cultural life" is more sociological. In fact, Robert E. Dickinson[27] has defined the scope of urban geography in terms very similar to those used by Burgess to define the "spatial pattern" aspect of urban sociology. For Dickinson, urban geography consists principally of the study of site and situation, and of the size, functions, spacing, layout, and build of cities. He goes on to point out that a considerable overlap exists between the work of urban sociologists and urban geographers, and that, especially, the spatial aspect of human ecology is "common ground" to both sociology and geography. Although one human ecologist has indicated that there is now a tendency for urban sociologists to "turn away from a concern with spatial patterns per se,"[28] there would appear to be little indication that urban sociologists plan a wholesale withdrawal, or propose abandoning the field to the urban geographers.[29] The problem of distinguishing the two centres on the fact that although some aspects of urban sociology can be separated fairly clearly from urban geography, there is great difficulty in clearly separating urban geography from urban sociology as a whole. For instance, the study of urban social problems such as family disorganization, organized crime, juvenile delinquency, gangs, narcotic addiction, loneliness and social isolation among the aged, etc. is clearly sociological and lies beyond the competence and interest of the urban geographer. On the other hand, there are no logical grounds on which to exclude the urban sociologist from the study of the spatial aspects of urban life, and in fact a number

---

[27] "The Scope and Status of Urban Geography: An Assessment," in Readings in Urban Geography, ed. by Harold M. Mayer and Clyde F. Kohn, pp. 10-26.

[28] Leo F. Schnore, "Geography and Human Ecology," Economic Geography, XXXVII, No. 3 (1961), p. 217.

[29] For instance, if one compares standard books of readings in urban geography (Mayer and Kohn, Readings in Urban Geography) and urban sociology (Burgess and Bogue, Urban Sociology; Paul K. Hatt and Albert J. Reiss, Jr., eds., Cities and Society: The Revised Reader in Urban Sociology), one becomes aware of a great similarity between the geography volume, and the "ecological" sections of the sociology volumes (Burgess and Bogue, Part I, "Urban Ecology and Demography"; Hatt and Reiss, sec. 4, "The Spatial and Temporal Patterns of Cities"). Of interest is the fact that the geography volume contains articles by sociologists, and the sociology volumes contain articles by geographers.

of urban sociologists would regard such studies as providing a necessary foundation for the more distinctively sociological studies. For instance, Burgess argues that although the recent tendency among urban sociologists has been to devote more attention to social problems as such, "the ecological aspect permeates and conditions all others, and the findings of the sociological studies will be strongly influenced by the degree to which the ecological conceptual system and the actual areas of the city are recognized in the assembling of data for the questions that are being raised."[30] If Burgess is correct, then it follows that urban geography is little more than a specialized branch or lower level form of urban sociology and at best can provide only preliminary data needed for the study of the more distinctively sociological problems. In fact, urban geography, as a distinct discipline, would then be limited to a consideration of site and topography, the purely physical aspects, of urban establishments. And moreover, in the absence of geographical work, urban sociologists have displayed little reluctance in carrying out preliminary studies respecting the topography and physical layout of cities.

When one considers the broader relations between human geography and sociology, one is still faced with the problem of distinguishing the two. On the one hand, human geography can apparently be defined so broadly as to encompass much of the study of "social systems" and man's "artificial environment,"[31] whereas, on the other, sociology can be defined so broadly as to encompass the study of land use and the development and use of natural resources,[32] matters that are customarily regarded as of primary geographical interest. William Bunge[33] makes the amazing statement that "if it were not for the existence of central place theory, it would not be possible to be so emphatic about the existence of a theoretical geography independent of any set of mother sciences," and then proceeds to discuss the views of a number of thinkers who were not geographers, nor

---

[30]"Research in Urban Society," 11.

[31]See, for instance, Philip Wagner, The Human Use of the Earth, especially chap. i, introduction, and chap. vii, "Artificial Environments."

[32]See, for instance, Walter Firey, Man, Mind and Land: A Theory of Resource Use.

[33]Theoretical Geography, p. 129.

so regarded themselves.[34] Yet, "central place theory" has been of some concern to sociologists on and off for at least the last seventy years.[35]

One of the major difficulties of drawing any very clear distinction between sociology and human geography is that both disciplines are defined so broadly that they include facets of all the other sciences that are usually classified as social or human. For instance, sociology is ordinarily divided into economic, political, urban, etc., just as human geography is ordinarily divided according to the same categories. We do not, however, ordinarily divide the more specific branches of the social sciences, such as economics and political science, according to the other social scientific categories. This situation gives rise to a tendency to absorb the whole of social science within either sociology or human geography. With respect to sociology, this is apparent in the writings of some of the greatest of sociological theorists. For instance, Emile Durkheim, one of the acknowledged founders of sociology, divides that discipline into three major categories: social morphology, social physiology, and general sociology.[36] Social morphology is concerned with the geographical basis of social life, and embraces such matters as the volume, density, and distribution of population, and in general, the whole study of the spatial aspects of society, matters which a number of

---

[34]The quotation given previously clearly indicates that Lösch regarded himself as an economist engaged in an important aspect of economic theory. Even Christaller, nominally a geographer, casts doubt on the geographical purity of his own work when he remarks: "Is what we have offered in the present study economic geography or national economics? . . . The main matter is the posing of the question, and the question is, without a doubt, geographical. The problems which are to be solved are geographical. They can be solved, however, only with the help of economic theory and economic methods" (Walter Christaller, Central Places in Southern Germany, p. 201).

[35]Charles Horton Cooley, "The Theory of Transportation," in his Sociological Theory and Social Research (1930), pp. 15-118; originally published, 1894. See especially preface (pp. 17-18), and sec. 1, "Mechanical and Geometrical Notions" (pp. 19-24), in which he discusses some of the basis of "central place theory," and evidently regards it as being of sociological concern. However, in his preface, Cooley refers to work being done in the field by geographers, notably to that of Ratzel. In addition, Edward Ullman, "A Theory of Location for Cities," in Readings in Urban Geography, ed. by Mayer and Kohn, p. 203, contains a considerable bibliography from the literature of rural sociology that bears on the question of "central place theory."

[36]"Sociology and Its Scientific Field," in Emile Durkheim, 1858-1917, ed. by Kurt H. Wolff, pp. 354-375.

geographers would regard as falling within their competence. Social physiology comprises the study of more distinctively sociological problems, yet is divided into categories such as sociology of religion, law, economics, etc., which cut across all the other recognized social or human sciences. General sociology consists of the search for general social laws, and is also a sort of social philosophy, defined so broadly as to amount to virtually a general theory of social science. Morris Ginsberg, the dean of British sociologists, discusses the scope of sociology in terms very similar to those employed by Durkheim.[37] Talcott Parsons, one of the leading theoretical sociologists in the United States, defines his theory of social systems so broadly that it encompasses economics and a variety of other social sciences. For instance, two of the general conclusions he draws from his study of the relations between the society and the economy read as follows: "Economic theory is a special case of the general theory of social systems and hence of the general theory of action"; "An economy, as the concept is usually formulated by economists, is a special type of social system."[38] On the other hand, especially in recent years, some geographers have defined the scope of human geography so broadly that it embraces much of social, economic, and political process.[39]

Nevertheless, despite a tendency to define sociology so broadly that it encompasses most of social science, we still recognize a number of problems as being peculiarly social or sociological and hence we can conceive of sociology as having a core. A partial list of such problems, relative to urban sociology, was given in an earlier context. Although I regard it as futile and even dangerous to attempt to draw absolutely rigid boundaries between sciences, nevertheless, if a discipline is to be recognized as independent, then it must possess a core of problems distinct from those studied by other disciplines. Does geography possess a core, and are there any distinctly geographical problems?

[37]Sociology, chap. i, "Scope and Method of Sociology," pp. 7-37.

[38]Talcott Parsons and Neil J. Smelser, Economy and Society: A Study of the Integration of Economic and Social Theory, p. 306.

[39]This tendency is quite evident in the Report of the Ad Hoc Committee on Geography, The Science of Geography. A discussion of this work will be considered in more detail in relation to contemporary concepts of the nature of geography.

Many geographers regard regional geography as the core of their discipline.[40] Although studies which can be labelled "regional economics" and "regional sociology," i.e., the study of economics and sociology in specific regional contexts, are clearly recognizable, regional geography, in one of its interpretations, aims at more. It attempts an integration or correlation, within regional contexts, of data that are studied, or even knowledge that is obtained, by a number of more specialized and precise disciplines. Yet, within recent years, a couple of "new" sciences, "ekistics" and "regional science," have arisen that have challenged the status of regional geography, and that appear to cover much the same ground that is encompassed within regional geography.

The scope of "ekistics" has been set forth recently in the following terms:

As ekistics treats the spatial organization of the environment from an interdisciplinary point of view, the subject matter throughout includes such aspects as the aesthetic, anthropological, economic, geographical, historical, political, social and technological.[41]

Ekistics appears to consist of a glorious spatial synthesis of virtually everything human, and also includes what is "geographical." In this context, "geographical" apparently refers to the study of man in relation to his physical environment, since geography, conceived as regional geography, would obviously cover at least a portion of the territory encompassed by the concept "ekistics."[42]

Walter Isard, the leader of the "regional science" movement, has defined the scope of regional science in the following terms:

[40]The following paragraphs should not be regarded as an adequate discussion of the nature of regional geography; rather, they consist only of a few preliminary considerations of that aspect of geography relative to some other "disciplines."

[41]Athens Centre of Ekistics, Athens Technological Institute, International Seminar on Ekistics and the Future of Human Settlements (Athens, Greece, July 20-24, 1965), pamphlet.

[42]However, in a different context, Doxiadis argues that the three sciences of terrestial space are geography, regional science, and ekistics. Geography is the most general yet the most factual of the three, since it deals with terrestial space as a whole. Regional science and ekistics are more limited in scope yet more theoretical than geography; regional science deals with "regions," and ekistics with "human settlements." Hence, in relation to geography, Doxiadis sees ekistics as theoretical urban geography, the latter being merely the empirical component of ekistics (Constantinos A. Doxiadis, "Ekistics and Regional Science," Regional Science Association, Papers, X [1963], 9-46).

. . . even more basic is the need for a much more comprehensive general
theory which not only covers equilibrium with respect to location, trade,
price, and production for a system of regions whose boundaries and trans-
port network are themselves variables, but also treats the fundamental
interaction of political, social, and economic forces as these interactions
affect the values of a society, condition its behavioral patterns and goal-
setting processes and lead to concrete decisions and policies relating to
interregional structure and function.[43]

Isard, perhaps, throws a little light on the matter in the following quotation:

Unless we do so [adopt the regional science approach], we shall probably
evolve regional analysis along traditional lines. We shall thereby be forced
to rely upon inferior additive processes to derive a total view of regional
structure, e.g., the addition of elements of regional economics, regional
sociology and regional geography. As a consequence, we shall fail to cap-
ture the essence of the region. More important, by adhering to a traditional,
outmoded slicing of the regional body, which yields the standard, social science fields of investigation, we may preclude the identification of the very
basic interaction matrix of this live, dynamic organism, which we seek.[44]

Although it is difficult to attach any precise meaning to what
Isard is saying, apparently his point of view involves a total
regional synthesis of everything human, including human values, the regional aspects of all social sciences, and also the
content of regional geography. However, he does point out
that this goal is likely to be achieved only "in the millennium"
(p. 20). Hence, total regional integration appears to be more
a scientific ideal than an immediate or even reasonable goal of
investigation. In the above quotations, Isard does not indicate
how he proposes to distinguish between regional science and
regional geography. This he does in the following remark: "I
view the regional scientist and geographer as general spatial
practitioners over the several social sciences, one at the abstract, theoretical level, the other at the more empirical,
grass roots level."[45] One could interpret this remark in several ways—for instance, that geography is little more than
empirical regional science, or that regional science provides
the theoretical basis for geography. However, any grounds

[43]Walter Isard et al., Methods of Regional Analysis: An Introduction to Regional
Science, p. xi.

[44]"Regional Science, the Concept of Region, and Regional Structure," PPRSA, II
(1956), 13.

[45]Isard, "The Scope and Nature of Regional Science," PPRSA, VI (1960), 15.

for drawing this kind of distinction have become seriously blurred, since in recent years, geographers, in increasing numbers, are turning in their work to the use of statistical and mathematical "models," and more "theoretical" approaches.

In addition, the whole idea that the social sciences require "spatial practitioners" may be questioned. It does not appear that the physical and biological sciences require such people from outside their own disciplines. Moreover, there is evidence to suggest that social scientists are perhaps doing more and more of their own "spatial practitioning."[46] What, if anything, then, is peculiar about the social sciences in this regard? Pitirim Sorokin[47] once drew a distinction between what he called the "horizontal" and the "vertical" dimensions of the "social universe." He was writing, to some extent, in opposition to the tendency of human ecologists to interpret and present all sociological matters in spatial terms. By the term "horizontal" Sorokin meant the spatial aspects of social life, and it included all social problems that could be defined and studied in spatial terms. However, Sorokin points out that many sociological questions, such as those pertaining to role, status, and family relations, cannot reasonably be interpreted and studied in spatial terms. These types of questions comprise the "vertical" dimension of sociology. Or, to put the matter in Kantian language, there is evidently an "inner" or non-spatial side to social science. Geographers have occasionally attempted to employ a distinction between "vertical" and "horizontal," although not very successfully. For instance, de Jong defines "vertical interrelation" as "the integration of things and their phenomena at any point on the earth surface."[48] But, if "vertical interrelation" occurs "at a point," it is obviously non-spatial, and one may wonder why it concerns the

[46]For instance, see Otis Dudley Duncan, Ray P. Cuzzort, and Beverly Duncan, Statistical Geography: Problems in Analyzing Areal Data, especially chap. i, "Preliminaries," pp. 3-31, which contains a considerable review of "spatial" literature from a variety of social sciences. Moreover, it is perhaps significant that this pioneer work in "statistical geography" was produced by a group of sociologists. Although the authors are well aware that they are doing something geographical in one sense of that term, they offer no apology; evidently they regard the pursuit of areal or spatial studies as sociologically legitimate.

[47]Social and Cultural Mobility, especially chap. i, "Social Space, Social Distance, and Social Position," pp. 3-10.

[48]Chorological Differentiation, p. 27.

geographer. However, de Jong indicates that "vertical inter-relation at a point" refers to "a small earth space" (pp. 29, 82). But "interrelation" within "a small earth space" is no longer non-spatial or "vertical" and hence must be spatial or "horizontal." In addition, since there are no grounds upon which to draw a hard and fast distinction between "a small earth space" and "a large earth space"—since at what point does "a small earth space" cease to be small?—there would appear to be no logical grounds whatever, in this context, upon which to draw a distinction between "vertical" and "horizontal." If Sorokin's distinction between "vertical" and "horizontal, " and Kant's distinction between "inner" and "outer, " can be main-tained, and I think they can, then we have grounds upon which to separate clearly some aspects of social science from human geography, if we assume that geography, by its very nature, is concerned, in some sense, with spatial matters. The difficulty, however, arises when one attempts to separate geography, con-ceived as a science of spatial relations, from the various social sciences, since there would appear to be no reasonable or log-ical grounds upon which to exclude the various social sciences from the study of spatial relations that concern them. The study of such matters is obviously and logically part of the content of social science once it passes beyond pure theory and the anal-ysis of elementary facts. Moreover, concern with the spatial and temporal dimensions of social scientific matters is nec-essary to secure a "fit" or a justification for at least some as-pects of social science theory. This view that geography, conceived as a set of individual specialties (economic geog-raphy, political geography, urban geography, etc.), is log-ically subsumable under the more theoretical disciplines (economics, political science, sociology, etc.) that stand in close relation to each of these geographical specialties, is by no means new or startling to geographers themselves. In fact, a number of geographers have argued for this view.[49]

However, despite the fact that regional science to date con-sists of little more than regional economics, Isard evidently re-

[49]For instance, see Nevin M. Fenneman, "The Circumference of Geography, " Annals, AAG, IX (1919), 3-11; Carl Sauer, "Foreword to Historical Geography, " in Leighly, ed., Land and Life, pp. 351-379; and Wooldridge and East, The Spirit and Purpose of Geography, chap. i, "The Nature and Development of Geography, " pp. 13-24.

gards both it and regional geography as being far wider in scope than the spatial interests of any one social science. Hence, regional science and regional geography are "general spatial practitioners, " and hope to achieve some sort of general spatial integration of data that cuts across all the recognized social sciences. But, does even this expedient secure an independent place for either regional science or regional geography relative to the whole of social science? Does it not merely shift the problem to another level? Just as the "spatial practitioners" are concerned with achieving a degree of spatial integration of data that are ordinarily studied by specific social sciences, so some social scientists themselves have long been interested in achieving a greater degree of theoretical integration among the several social sciences. For instance, Talcott Parsons and the group of social scientists around him at Harvard University have for some time now been interested in a theoretical integration of some aspects of economics, sociology, social psychology, and anthropology, under the name of a "general theory of social systems. " If Parsons and his co-workers succeed in creating a general theory of social systems that possesses empirical warranty, will they not be just as interested in the spatial and temporal dimensions of that theory as practitioners in the particular social sciences are today interested in the spatial and temporal dimensions of their more specific theories? And moreover, will not concern with the spatial and temporal dimensions of the general theory of social systems be presupposed as necessary to secure a "fit" or a justification, or empirical warranty, for that theory?

Is it perhaps possible to draw a distinction between geography in the broad sense and geography in the narrow sense, just as it seems possible to draw a distinction between history in the broad sense and history in the narrow sense? I understand "history in the narrow sense" to encompass the primary concerns of the professional historian, whereas "history in the broad sense" can be thought of as falling more readily within the competence of a variety of scientists insofar as they are concerned with the historical dimension of their respective disciplines. For instance, Max Weber once attempted to draw a distinction between sociology and history in the following terms:

176

It has continually been assumed as obvious that the science of sociology seeks to formulate type concepts and generalized uniformities of empirical process. This distinguishes it from history, which is oriented to the causal analysis and explanation of individual actions, structures, and personalities possessing cultural significance. The empirical material which underlies the concepts of sociology consists to a very large extent, though by no means exclusively, of the same concrete processes of action which are dealt with by historians.[50]

Weber's approach to sociology was often an historical one, and he sought to secure its independence from history on the grounds that sociology is concerned with typical situations and with formulating "generalized uniformities," whereas history is concerned with the study of the "individual." Individuals, understood in the sense of particulars, would indicate that history studies the same matters as does sociology, though at a concrete or non-general level. But individuals understood in the sense of the unique or the non-repeatable would indicate a concern with matters that are not encompassed even as particular cases within the "generalized uniformities" that sociology seeks to formulate. This issue of the relations between history and other sciences would appear to be confined to the social or human sciences. For instance, the theory of evolution is historical biology, but historical biology, even in an "individual" sense, is in no sense the concern of the professional historian. The issue is not so simple with respect to geography, since many geographers are still concerned with physical and biological matters. Yet, it might be possible to draw a distinction between geography and other sciences on the grounds that geography does not seek to formulate "generalized uniformities," but rather confines itself to the study of the "individual," in the sense of either the particular or the unique and non-repeatable.

What I think has emerged from the foregoing survey of literature from a number of fields bearing some relation to geography is that both geographers and outsiders distinguish geography from other sciences on the basis of a variety of concepts. Let us take a closer look at these various concepts of the nature of geography.

---

[50]Max Weber, The Theory of Social and Economic Organization, p. 99.

VIII
Contemporary concepts of the nature of geography

Six contemporary concepts of the nature of geography can be fairly clearly distinguished from each other. These may be briefly labelled: (1) environmentalism and ecology; (2) spatial relations and spatial distributions; (3) areal differentiation and areal integration; (4) place and location; (5) synoptic; and (6) applied. These concepts are not necessarily mutually exclusive, and they may to some degree overlap with one another. In addition, there is nothing especially "contemporary" about any of them, since all six are to be found, to a greater or less extent, in Kant's concept of geography.

The following analysis of these various concepts of the nature of geography will attempt, essentially, to explicate the meaning of each, and comparatively the merits of each, as providing an adequate logical foundation for geography as a distinct scientific discipline. In each case, comparison will also be made with Kant's concept of geography, so as to gain further insight into that conception of the discipline, and into its possibilities of providing a foundation for geography. References will be made to the ideas of a number of contemporary geographers, although these ideas are included primarily for illustrative purposes. Hence, no pretense is made to exhaust the number of proponents of any point of view discussed, although it is intended that the persons mentioned are of some consequence in current debate respecting the nature of geography.

1. ENVIRONMENTALISM AND ECOLOGY

a. Environmentalism

By the concept "environmentalism" I mean the study of man in relation to his physical environment.[1] Within this broad context, the concept has been subject to a variety of interpretations. It is doubtful, however, that any twentieth-century geographer has held the doctrine in the extreme seventeenth and eighteenth-century form of "environmental determinism" that was so convincingly refuted by Hume. Even when contemporary geographers, such as Griffith Taylor, have employed the concept "environmental determinism," the "determinism" part of the concept has been used in an odd sense. Ordinarily, geographers think of determinism in connection with the limited sense of the determinism versus free-will controversy, and not in connection with the metaphysical doctrine that everything that happens in the universe is governed entirely by laws. Yet, no precise meaning can be attached to the idea that the physical environment determines my mental processes so that I choose to do one thing rather than another. A moment's reflection is all that is necessary to convince us that the physical environment is one of the conditions of much of human behaviour. Even in an advanced industrial civilization, some facets of our behaviour are different in summer than in winter, and different on sunny days than on rainy days. However, the physical environment did not determine that men should build large cities in

---

[1]Although geographers have often used the concept "environmentalism" as a synonym for "environmental determinism," it seems preferable to use it as a convenient label to refer to any study of man in relation to his physical environment. For comparable usage and discussion of this point, see Gordon R. Lewthwaite, "Environmentalism and Determinism: A Search for Clarification," Annals, AAG, LVI, No. 1 (1966), 1-23. A considerable body of geographical literature exists on "environmentalism." Hence, only a few issues of central philosophical importance will be examined here. Aside from Lewthwaite's article, perhaps the most comprehensive recent treatments of the topic are to be found in Harold and Margaret Sprout, The Ecological Perspective on Human Affairs; and in Tatham, "Environmentalism and Possibilism."

which to live, nor that they should invent automobiles that require antifreeze and get stuck on snowy winter days, nor that I
should choose to carry an umbrella or not when it rains.
Morris Cohen[2] has pointed out that an environmental determinism of this type leads to an obvious contradiction, since throughout history men have repeatedly chosen to leave inhospitable
environments in search for ones they regarded as better; yet
many people choose to remain in inhospitable environments.
Hence, environmental determinism supports two contradictory
propositions: men choose to leave because of the environment;
men choose to stay because of the environment. In fact, Griffith Taylor himself rejected "the effect of environment on in-
tangibles such as character and temperament,"[3] and presumably on specific mental processes, as being unscientific. His
"scientific determinism" consisted principally of the study of
the limits of human settlement. Taylor talks of a "plan of nature" which is beyond human control, yet which men must learn
to comprehend and live in accordance with.[4] Since men do not
as yet understand "nature's plan," they may make any number
of foolish decisions respecting the uses to which they put the
earth's surface. Yet, since men are apparently free to choose
between acting wisely and acting foolishly, it is difficult to see
in what sense Taylor's "scientific determinism" is any longer
a variety of determinism, or differs substantially from "possibilism," the doctrine which states that although the environment always imposes limits at any given time on what man can
accomplish, he is free within those limits to make any number
of rational choices respecting his use of the earth's surface.[5]
Taylor's "environmental determinism," however, appears to
culminate in a speculative philosophy of history of the "determinist" variety; he seems to have believed that "scientific determinism" will some day triumph and hence man will eventually learn to live rationally in accordance with "nature's plan."

[2]Reason and Nature, p. 343.

[3]Editor's introduction, "The Scope of the Volume," Geography in the Twentieth Century, p. 7.

[4]Ibid., "Editor's Note," pp. 161n-162n.

[5]George Tatham has pointed out that Griffith Taylor's "stop-and-go" determinism is
philosophically indistinguishable from "possibilism" (ibid., pp. 159-161).

Numerous twentieth-century technological developments
have served to dull the earlier impact of ideas respecting plans
and limits of nature. In addition, these technological advances
mean that the limits of the environment are steadily being
pushed back, although it is reasonable to assume that there are
limits as to what any stage of technological development can
accomplish. There are also ultimate environmental limits;
e.g., there are undoubtedly ultimate limits, independent of
technology, respecting how many people can be supported on
the planet Earth, since that planet is of a relatively fixed and
finite size. Because of a decline in the force of ideas concerning plans and limits of nature, contemporary geographers
interested in environmental studies tend to stress man's role
in relation to nature, rather than nature's influence on man.
Hence, such studies stress how man has used the surface of
the earth, how he is now using it, and how he can and ought to
use it. [6]

This tendency to be more concerned than formerly with man
as actively engaged in changing nature or the environment has
given rise in recent years to a variety of geographical literature
that has come to be known as "perception" studies. [7] The term
"perception" is used here more broadly and loosely than it
would ordinarily be employed in contemporary philosophy or
psychology, and appears to include conception, reasoning,
imagination, and memory, as well as the apprehension of ordinary sense-objects. Nevertheless, this type of literature
indicates a growing awareness that the environment in its relations to man cannot profitably be regarded as altogether unitary,
that what is environment may indicate different things to different men in the same or different situations, and that men
may see their situations relative to the environment in many
different ways. Perhaps the best statement of this orientation
in environmental literature is still that of Erich Zimmermann. [8]
Nature is "neutral stuff" until man sees some aspect of it as

[6]Various examples of this type of literature are to be found in William L. Thomas,
Jr., ed., Man's Role in Changing the Face of the Earth.

[7]A recent example of this type of literature is Ian Burton and Robert W. Kates, "The
Perception of Natural Hazards in Resource Management," Natural Resources Journal,
III, No. 3 (1964), 412-441.

[8]World Resources and Industries, especially chap. i, "Meaning and Nature of Resources," pp. 3-20.

fulfilling certain of his needs, as performing certain functions for him, or as fitting in with certain other of his plans; a natural resource is not a resource at all until man sees it in some relation of means and ends to himself. Moreover, this orientation suggests a change in philosophical basis towards a phenomenological one. Zimmermann's discussion is reminiscent of Sartre's distinction between "environment" and "situation." "Environment" is simply what it is, "neutral stuff," something that surrounds or environs us and is "out there," something that is only generally and peripherally a part of human consciousness, whereas "situation" is "environment" made meaningful because of some specific human intention. By way of illustration, before Saint Bruno and his followers arrived in the French Alps in 1084, the mountains were not beautiful, or awe-inspiring, or majestic, or forbidding, they were simply "environment," something that could only be described in the language of physical geography. But in seeking a hermitage, to Saint Bruno and his followers, the mountains ceased to be "environment" and entered into a "situation" with them. The mountains now became a retreat, safe or dangerous, suitable or unsuitable. Once a suitable place for the hermitage had been located, this particular dialectic between man and the "environment" ceased, and the mountains again became simply "environment." But immediately, it was necessary to secure building materials for shelter. At once, the mountain forests ceased to be "environment," and entered into a new "situation" with man in which they became rich or poor in wood, contained useful or useless wood, construction lumber or firewood. One can imagine this dialectic between man and "environment" continuing down to the present day, and giving rise to a variety of "situations."[9]

There is little in the foregoing, aside from the phenomenological point of view, that is not to be found, at least inchoately, in Kant's concept of geography. In his earlier days, Kant embraced a form of "environmental determinism" very similar to that of Montesquieu, yet, as we have seen, he included man as one of the principal agents that bring about changes in the earth's surface. For Kant this process was ev-

[9]The reference to Sartre and the illustration are taken from Stephan Strasser, Phenomenology and the Human Sciences, pp. 282-283.

idently two-way; if it is true that the environment conditions human activity to a degree, it is equally true that man changes that environment to a degree. The environment can hardly be said to determine man to change it in determinate ways. With respect to the phenomenological point of view, so far as I have been able to ascertain, Kant does not discuss it explicitly. Yet it would appear to be implicit in some aspects of his teaching on geography, although I undoubtedly run the risk of reading more into Kant than is really there. At the empirical level, nature does not exert an inherently necessary effect on human mental processes, and so man is to some extent free to view nature according to his individual purposes. In addition, nature in the empirical sense is not something devoid of ordinary human interest, since nature, understood in the geographical sense of the surface of the earth, is the "home of man," and hence the more man knows about his physical environment the better able he is to live in it and cope with it. But since geographical knowledge is limited primarily to what can be directly experienced, then geography for me personally is limited to that portion of the earth's surface that I can directly experience, and in this way my personal physical environment is limited. An experientially based geography of the whole of the earth's surface requires an act of faith on my part. In order to arrive at this experientially based concept, I must accept the accounts of the experiences of others as being as accurate as my own.[10] Yet, since nature does not wholly determine my conscious mental processes, what I choose to see as my environment is at least to some extent the result of my own intentions.

b. Ecology

The concept "ecology" has never been very popular among the majority of geographers. Nevertheless, environmental studies are "ecological" in one sense of that term. Although this concept has been used in an almost bewildering number of senses and contexts,[11] the general account of the concept set forth by

[10]Kant, of course, as we have seen, does not strictly hold that an experientially based concept of the whole of the earth's surface is possible, since for him the concept "whole of the earth's surface" is an Idea.

[11]For reviews of a variety of senses in which the concept "ecology" has been used, principally in social science literature, see Emma C. Llewellyn and Audrey Hawthorn, "Human Ecology," in Twentieth Century Sociology, ed. by Georges Gurvitch and

its begetter, Ernst Haeckel, remains one of the best brief treatments. After drawing a distinction between the "inner" and "outer" aspects of biological organisms, i.e., "the functioning of the organism in itself," as contrasted with "its relationships with the outer world," Haeckel goes on to remark:

The outer physiology, the study of the relations of animals with the outside world, may in turn be divided into two parts, the ecology and the chorology, of animals. By ecology, we understand the study of the economy, of the household, of animal organisms. This includes the relationships of animals with both the inorganic and the organic environments, above all the beneficial and inimical relations with other animals and plants, whether direct or indirect.[12]

"Chorology" refers to the distribution of biological organisms over the surface of the earth. "Ecology" evidently has two aspects: (1) the relations of plants and animals to the physical or inorganic environment in which they live; and (2) the relations of plants and animals to other species that live in mutual association with them. Although Haeckel obviously intended that the second type of study comprise the main focus of ecology, Charles Singer[13] points out that both types of study have frequently been labelled "ecology." When the term passed into social science literature as "human ecology," principally through the work of Robert Park, it was used primarily in the second sense, to refer to the spatial relations of men and their various institutions and activities. Because of man's great mobility and relative freedom from specific environmental controls, a scientific human ecology is much more difficult to achieve than a scientific plant ecology, and in fact the whole idea of a human ecology may be dismissed as a vague biological analogy.[14] If the concept "human ecology" is meaningful

Wilbert E. Moore, pp. 466–499; and Marston Bates, "Human Ecology," in Anthropology Today: Selections, ed. by Sol Tax, pp. 222-235.

[12]Ernst Haeckel, "Ueber Entwickelungsgang und Aufgabe der Zoologie," Jenäische Zeitschrift für Medicin und Naturwissenschaft, V (1870), 353-370; quoted in Bates, "Human Ecology," p. 222.

[13]A History of Biology, pp. 283-284.

[14]For instance, Gideon Sjoberg has recently attacked as "of little value" the "biotic" point of view in human ecology, championed by Robert Park and his followers, i.e., that the city can profitably be studied as a "natural environment." Undoubtedly, part of the impact of labelling this viewpoint "biotic" is the inference that the analogy with plant ecology had been drawn much too literally. Yet, Sjoberg's "more adequate theory," that combines as variables "culture," "values," "technology," and "power," as well as the "biotic," is still regarded by the author as "ecological" (Sjoberg, "Comparative Urban Sociology," in Sociology Today: Problems and Prospects, ed. by Robert K. Merton, Leonard Broom, and Leonard S. Cottrell, Jr., pp. 334-359).

(and it appears to be well entrenched at present in social science literature), then ample precedent exists in biological literature for extending the concept to include the relations between man and his physical or inorganic environment.[15] In fact, at least one geographer has proposed that the concept be defined in just this sense:

> The center of gravity within the geographic field has shifted steadily from the extreme physical side toward the human side, until geographers in increasing numbers define their subject as dealing solely with the relations between man and his natural environment. Thus defined, geography is the science of human ecology. The implications of the term "human ecology" make evident at once what I believe will be in the future the objective of geographic inquiry. Geography will aim to make clear the relationships existing between natural environments and the distribution and activities of man.[16]

For two principal reasons, the label "human ecology" never caught on as a substitute for "geography": (1) many geographers already regarded environmental studies as the central focus of their discipline, and hence were little inclined to change the long-standing name of the discipline; (2) the label "human ecology" has been usurped by sociologists, and hence the application of the same label to geography could produce only further confusion respecting grounds for a distinction between sociology and human geography.

c. Summary
Within recent years, there has been a tendency on the part of some geographers to dismiss environmental studies as of little consequence, and certainly as not comprising the central focus of the discipline. For instance, Hartshorne[17] has argued that the whole idea of environmental studies introduces a dualism into geography—a dualism which presupposes the existence of a gap that must be bridged between the physical or natural sciences and the human sciences.[18] Rather than being split into

[15]Geography as "human ecology" in the former sense appears to be indistinguishable from geography as the science of spatial relations. Hence, this topic can be left for consideration until we discuss that concept of the nature of geography.

[16]Harlan H. Barrows, "Geography as Human Ecology," Annals, AAG, XIII, No. 1 (1923), 1-14; quote from p. 4.

[17]Perspective on the Nature of Geography, chap. vii, "The Division of Geography by Topical Fields—The Dualism of Physical and Human Geography."

[18]Moreover, Hartshorne implies that if geography is conceived as the science that bridges the gap between the physical and human sciences, and if this task is to be accomplished in adequate fashion, then an enormous knowledge of both the physical and human sciences, beyond the competence of any one geographer, is presupposed.

physical and human parts, and seeking to bridge the gap between them, geography can only attain status as a unitary discipline by studying the partial or even total "integration" of data, irrespective of physical or human origin, within regional contexts. Furthermore, the present tendency of a number of geographers to regard their discipline as being primarily concerned with spatial relations and spatial distributions largely precludes interest in the relations between man and his physical environment. Nevertheless, as references given in the previous chapter indicate, a number of outsiders regard the study of man in relation to his physical environment as falling outside the scope of their particular disciplines, and as coming primarily within the competence of the geographer. For instance, scientists who are primarily concerned with the physical earth are rarely concerned with the implications of their studies for human life. Although biologists are intensely interested in the environmental aspects of plant and animal life, they have shown little inclination to be similarly interested in the relations between man and his physical environment. And social scientists, by and large, have eschewed the detailed study of the subject, and have left it to the geographer. Hence, it would appear that environmental studies comprise a research area that falls between the primary interests of the physical and biological sciences on the one hand and the social sciences on the other hand, and that is primarily geographical. Environmental studies can be regarded, therefore, as providing a reasonably distinctive core of problems for geography. Furthermore, this is an important research area, and if geographers abandon it, it seems unlikely at present that many other sciences will be inclined to claim it and research it, except peripherally.

This justification of a core for geography is, however, pragmatic and not logical. Since biologists are vitally interested in the relations between plants and animals and their physical environments, there would appear to be no inherent reason logically restricting social scientists from being essentially concerned with the study of man in relation to his physical environment. Nor has the present situation respecting this matter always been the case. For instance, Aristotle did not recognize the study of man in relation to his physical environment as the concern of a special science, but included it as a part of "politics."

2. SPATIAL RELATIONS AND
   SPATIAL DISTRIBUTIONS

More was said in the preceding chapter on this concept of the
nature of geography than on any other, primarily because it is
one of the most difficult to maintain if one is concerned with
securing an independent place for geography relative to other
sciences. Details from the preceding chapter need not be repeated; rather, the discussion may proceed at a more abstract
level. Since many geographers would agree that their discipline is concerned, in some sense, with "spatial relations,"
what I now have in mind respecting this concept is the view that
geography is the generalizing or law-finding science of spatial
relations occurring among phenomena on the earth's surface.
Representatives of this point of view would include Fred
Schaefer,[19] William Bunge,[20] and apparently the authors of the
Report of the Ad Hoc Committee on Geography, The Science of
Geography, who lay great stress on the "theoretical-deductive
approach." Basic to this concept of geography as a generalizing science of spatial relations is the idea that there are "three
great parameters of concern to scientists, space, time, and
composition of matter." Of these three "parameters," as they
apply to the surface of the earth, space is assigned to geography.[21] Or, to put the issue somewhat differently, unlike
other sciences, except possibly history, geography is not a
"subject matter" science, since it has no "matter" or objects,
the analysis and study of which is its special concern. However, in one respect, this is an odd distinction to make, since
one of the essential things we mean by the concept "science"
is some discipline that studies "matter" or objects in space
and in time. Basic to Kant's whole concept of science is that

---

[19]"To explain the phenomena one has described means always to recognize them as
instances of laws. . . . Hence geography had to be conceived as the science concerned
with the formulation of the laws governing the spatial distribution of certain features
on the surface of the earth" ("Exceptionalism in Geography," 227).

[20]See especially Bunge's discussion of "The Predictability of Geographic Phenomena,"
and his critique of the place of the unique in geography (Theoretical Geography, pp. 7-
13).

[21]The Science of Geography, p. 1. Strictly speaking, in this work, "space in time"
is assigned to geography. However, the additional assignment of time to the geographer serves only to complicate the basic issue. Discussion of this complexity is
reserved until later.

space and time are the a priori conditions of the possibility of experience, and hence of science. In other words, science studies objects in space and in time; a non-spatial, non-temporal science is both a perceptual and conceptual impossibility. Furthermore, since any science is a system, it studies relations among phenomena—relations that are basically spatial and temporal, regardless of whatever other labels we may apply to specific sets of relations. Failure to keep in mind these basic ideas respecting what is meant by science has led to some misleading statements in geographical literature concerning the nature of other surface-of-the-earth sciences. For instance, Schaefer remarks that

geography, thus, must pay attention to the spatial arrangement of the phenomena in an area and not so much to the phenomena themselves. Spatial relations are the ones that matter in geography, and no others. Nonspatial relations found among the phenomena in an area are the subject matter of other specialists such as the geologist, anthropologist, or economist. [22]

If we extend Schaefer's argument to include time, and assign the study of temporal sequences or relations to the historian, then the only conclusion respecting this matter that can be drawn is that economic, social, political, and other relations must be non-spatial and non-temporal. Hence, economics, sociology, political science, etc. are non-spatial, non-temporal sciences. But this is absurd, since no meaning whatever can be attached to the idea of a non-spatial, non-temporal science. When we employ a phrase such as "economic relations," we are using a convenient shorthand that stands for "economic spatio-temporal relations." And insofar as economics qualifies as a science possessing empirical warranty, then its generalizations must apply to given spatio-temporal situations.

Since all sciences study spatial relations, five possible alternatives appear to be open respecting the concept of geography as a generalizing science of spatial relations: (a) geography studies geographical spatial relations as distinct from economic, social, and other spatial relations; (b) geography is a super-science of spatial relations; (c) geography is a lower-order or lower level science of spatial relations, studying economic, social, and other relations at a more concrete or

[22]"Exceptionalism in Geography," 228.

particular level than that of various other sciences to which it is closely related; (d) geography studies the spatial relations among "things in reality," rather than in terms of sets of abstractions such as economic, social, political, etc.; (e) geography is not a generalizing science of spatial relations at all, but rather a different interpretation of the concept "spatial relations" is required.

a. Geography as the Study of
   Geographical Spatial Relations

The difficulty of maintaining the thesis that there is a special set of geographical spatial relations left over after all other sciences have removed their particular sets of spatial relations from the spatial pot stems from the fact that geography apparently has no special set of objects that is peculiarly its own. If geography has no objects of its own, it is difficult to see how it can have a set of spatial relations that is peculiarly its own. However, when geographers talk about their discipline as being a generalizing science of spatial relations they appear to have one or more of three ideas in mind: (1) geography studies the spatial distribution of phenomena over the surface of the earth; or (2) geography is the science of "spatial interaction"; or (3) geography is the science of "spatial integration."[23] With respect to the ideas of spatial distribution and "spatial interaction," we undoubtedly have something different in mind than when we maintain, as an initial condition of all sciences, that they must study things in space and time. In the latter case, when it is maintained that scientific generalizations must apply in space and in time, we have in mind applicability in the sense of spatio-temporal coordination, i.e., specific location in space and in time, and not necessarily in the senses of spatial distribution, spatial interaction, or changes through time.

---

[23]Geographers do not always clearly distinguish between "spatial interaction" and "spatial integration." I understand the concept "spatial interaction" to refer to the mutual conditioning relations which may exist among phenomena that are subsumable under a single recognizable class, e.g., physical, biological, economic, social, etc. On the other hand, the concept "spatial integration," although it presupposes "spatial interaction," implies much more—it implies a partial or even total correlation of data that are studied by any number of specialized disciplines. Hence, a consideration of "spatial integration" can best be left until later, when we consider other interpretations of geography as a science of spatial relations.

The threefold distinction that Kant draws between spatial relations among phenomena on the earth's surface as they occur in
the present, and in the past, and changes in phenomena over
time, is undoubtedly legitimate. But it does not follow that
there can be such a thing as a purely spatial science, or a
purely temporal (natural) science, or that spatio-temporal coordination has been abandoned. Kant is quite clear about the
fact that although the focus of a natural historical study may be
the development of things through time, the things studied are
referable to specific spatial locations. And even if the historian adopts the approach of being primarily concerned with understanding the "inside" of historical occurrences (to adopt
Collingwood's term), i.e., attempts to put himself in the shoes
of the historical actor and to rethink the thoughts that led him
to act as he did, an understanding of the "inside" of historical
occurrences implies at the same time an understanding of the
"outside" of such occurrences. That is, the historian often
has to begin with the "outside of the event, " with the event as
a physical occurrence that involves spatial relations among
phenomena.[24] Again, Kant is clear about the fact that although
the focus of a geographical study may be the spatial association
among things, those things are referable to a definite point in
time. And insofar as geography for Kant is concerned with
spatial interaction—as it would appear to be, since as an empirical science it must assume the postulate of empirical
thought that he calls the "principle of community"—since spatial interaction implies contact or influence over distance, and
since such interaction requires time in which to happen, then
a time factor is required and is inevitably built into the considerations. The same argument applies to the employment of
causality in geography, since a temporal sequence is involved
in passing from cause to effect. There is, however, undoubtedly a sense in which a distinction can be drawn among spatial
distributions and interactions, changes through time, and the
analysis of the "composition of matter. " For instance, we
ordinarily think of the botanist as being primarily concerned
with the analysis of the properties of various plants. Yet, concern with the analysis of the "composition of matter" does not
preclude spatio-temporal coordination or spatial relations.
When a botanist analyzes a plant into various "parts, " the

[24]R. G. Collingwood, The Idea of History, pp. 213 ff.

"parts" still stand in spatial association to one another; when he is concerned with the growth of a plant, he is concerned with developmental changes that proceed over time; and in each case, the plant and its "parts" can be coordinated spatially and temporally.

However, despite the fact that the foregoing distinctions can legitimately be made, they do not provide us with any grounds for recognizing a special set of geographical spatial relations. As indicated in the preceding chapter, at the macro-level, the study of the distribution of phenomena over the surface of the earth, and of the spatial interaction of such phenomena, would appear to be a legitimate and obvious part of the content of the physical and biological sciences. Nevertheless, the issue appears to be less clear-cut respecting the social sciences. Is this because the social sciences are less developed? And does not increasing maturity among the social sciences imply, as Alfred Weber suggests, growing concern with the spatial interaction and distribution, and temporal development, aspects of social scientific matters?

We appear to be driven back to reconsider what is meant by saying that a science studies a certain set of objects. To begin with, it is evident that any science studies things in space and in time. In other words, it is just as impossible to conceive of a pure science of space per se or a pure science of time per se, except possibly in the sense of pure mathematics, a geometry or an algebra, as it is to conceive of a science that studies things non-spatially and non-temporally.[25] A common way of thinking about the divisions and even sub-divisions among sciences is in terms of differences in the kinds of objects they study. But this way of thinking about divisions among sciences appears to fit some sciences better than it does others. For instance, we think of the botanist as studying plants in all their ramifications, the zoologist as studying animals, the geologist rocks, and the pedologist soils. "Universal" sciences such as physics and chemistry, however, are more difficult to classify in this way, since they appear to cut across the elementary distinctions that can be made among different kinds of discrete objects. For instance, men, animals, and plants, as well as

---

[25]When Bunge (Theoretical Geography, p. 197) indicates that geography is concerned essentially with the study of "spatial problems per se, " it is difficult to see how that discipline can be anything more than a branch of mathematics.

inanimate objects, are all subject to a variety of general physical laws, and in this sense can be thought of as objects for physics; the chemist studies chemical elements, yet men, animals, and plants, as well as inanimate objects, are composed of these elements.

The social sciences present even greater difficulties in this regard. Collectively, we can think of them as studying man in a social sense, i.e., human relations, activities, institutions, etc. Yet, taken individually, it is difficult to see them as studying objects in quite the same sense as the botanist studies plants or the geologist studies rocks. It seems more reasonable to regard them individually as studying certain sets of relations that have been abstracted from the collective or general object of study, man as a social animal. Nevertheless, spheres or areas of predominant interest and concentration can be recognized among the various social sciences. For instance, the political scientist studies government in all its ramifications, the economist studies monetary and business matters, and the sociologist focusses his attention primarily on a number of distinctively social problems and institutions. History and geography present further difficulties in this regard. It is often argued that they are not "subject matter" sciences, that is, they have no special sets of objects the analysis and study of which is their special concern. However, it does not follow that either history or geography is not concerned with the study of objects in some sense. If history and geography have no objects of their own, then they must study the same objects as do a variety of other disciplines. However, if they are to be maintained as independent disciplines, then they must study objects either at a different level than do other sciences, or fundamentally different perspectives must be involved in history and geography. The idea of a "different level" introduces the second and third possible interpretations of the concept of geography as a generalizing or law-finding science of spatial relations.

b. Geography as the Super-
   Science of Spatial Relations
Although no contemporary geographer explicitly holds the view that geography is a super-science of spatial relations, there

has been a tendency towards such a position on the part of some geographers ever since Ackerman announced that "study of the evolution of space content on the earth's surface is the fundamental research frontier."[26] "Study of the evolution of space content" gives geography an enormous scope, since "space content" includes not only everything we can observe on the earth's surface, but also, for Ackerman, more intangible things such as "cultural processes" which include population movement and human needs, political organization and administration, and economic and technological development (pp. 24-26). However, in fairness to Ackerman, it should be pointed out that his concern with the "evolution of space content" is limited to considerations of whatever brings about changes in space relations and spatial distributions. Nevertheless, even this limitation still leaves geography with an enormous scope and content, and also re-raises basic logical issues respecting the concepts "space," "time," the "analysis of the composition of matter or objects," and the relations between them. This tendency towards super-science status for geography has recently been carried to further extremes in the Report of the Ad Hoc Committee on Geography, The Science of Geography, where, as remarked earlier, "space in time" is assigned to geography, and all other surface-of-the-earth sciences are presumably limited to considerations of the "composition of matter." A fundamental statement of this position may be seen in the following remark:

Geographers believe that correlations of spatial distributions, considered both statistically and dynamically, may be the most ready keys to understanding existing or developing life systems, social systems, or environmental changes. They further believe that geography has made a significant contribution in the past to the foundations of knowledge needed to understand subsystems of the man-environment system.[27]

How is one to interpret the idea that geography provides the "keys to understanding existing or developing life systems [and] social systems"? The concept "keys to understanding" is not spelled out. Are we to assume that geography provides the explanatory basis for biology and the social sciences, or

---

[26] Edward A. Ackerman, Geography as a Fundamental Research Discipline, p. 28.
[27] Report of the Ad Hoc Committee on Geography, The Science of Geography, p. 9.

193

that these sciences are somehow reducible to geography? Is some form of geographical or environmental determinism being reintroduced? Are we to assume that all considerations of spatial relations, spatial distributions, and developmental changes that effect such relations and distributions, are the exclusive prerogative of the geographer? Is history, and the historical dimensions of the biological and social sciences, also being reduced to geography? Geographers are apparently as fundamentally concerned with understanding the "great man-environment system," as they are with understanding spatial distributions over the surface of the earth, and spatial relations occurring among phenomena on the earth's surface. How is one to reconcile these apparently disparate concepts of what geography is fundamentally all about?[28] And finally, if geography is not at all concerned with the analysis of the composition of "matter" or objects, can one meaningfully talk about developmental changes that occur in purely spatial patterns, distributions, or interactions? For do not changes in the "composition of matter" and in the objects that various sciences study imply in turn profound changes in the spatial and temporal, or in the geographical and historical, dimensions of "matter" or objects? None of this is ever made explicit.

The basic logical issues involved here have been discussed earlier. Fundamentally, the issue reduces to the following— if geography is conceived as a generalizing science of spatial relations, interactions, and distributions, including the temporal or developmental aspects of such relations, interactions, and distributions, then the various biological and social sciences are destined to remain in a peculiarly truncated state, limited either to the study of highly theoretical or abstract matters devoid of empirical warranty or to the study of elementary analytical facts that are without broader spatial and temporal, or geographical and historical, dimension. In addition, since to date we have had little success in formulating generalizations or laws in the social sciences that possess broad empirical coverage or warranty, then the likelihood of estab-

---

[28]Edward Ackerman sheds some light on this issue in his article, "Where Is a Research Frontier?" Annals, AAG, LIII, No. 4 (1963), 429-440. In that article, it appears that the "great man-environment system" is the collective concern and "overriding problem" of all surface-of-the-earth sciences, whereas the study of spatial relations is geography's area of concentration within earth science as a whole.

lishing even broader geographical generalizations or laws that manage to sum up, spatially and temporally, a number of the findings of various social sciences would appear to be remote. In short, the issue of the conception of geography as a generalizing or law-finding science that somehow stands above the social sciences and history is not even appropriately debatable.

Furthermore, this recent tendency towards the reconstitution of geography as a sort of general or higher-order social science is reminiscent of the attempt in the last century by Gerland to revitalize geography as general geophysics, or as a general physical science of the planet Earth.[29] The latter attempt failed, not merely because outsiders refused to see geography's role in such a way, but also because many geographers themselves were little inclined to grant their discipline super-science status. The more recent attempt, although confined primarily to the social sciences, would appear to be equally destined to fail.

c. Geography as the Lower-Order or
   Lower Level[30] Science of Spatial Relations

Whether or not geography can be conceived as a lower-order or lower level science of spatial relations depends primarily upon whether or not a meaningful distinction, in terms of different classes of sciences, can be drawn between the general and the particular. In fact, however, a distinction in terms of levels of generality at which sciences typically operate, despite no essential difference in subject matter, is often made. For instance, the following argument is employed to distinguish micrometeorology and microclimatology:

Depending on the purpose of the investigation, it is useful to make a distinction between microclimatologic and micrometeorologic measurements, although in practice there are many intermediate stages.

Microclimatologic measurements should establish characteristic values of the meteorologic parameters for a particular place. Such values may be means, extremes, frequencies, duration or time of occurrence of certain values, and so on. . . .

[29]For discussion of Gerland, see Hartshorne, The Nature of Geography, pp. 102-120.

[30]The terms "higher-order" and "lower-order" designate a difference of level within a hierarchy, whereas the terms "higher level" and "lower level," although designating a difference of level, should not imply a hierarchy, i.e., deducibility of "lower" statements from "higher" ones.

Micrometeorologic research is differentiated from microclimatologic by the method used to analyze the recorded data. The primary data are used here to derive combinations of values of meteorologic parameters that will serve to establish or to prove meteorologic laws of general validity (that is, independent of locality).[31]

The aim of each science is different. Micrometeorology aims at the establishment of general laws that apply independently of any specific locality. Microclimatology, on the other hand, is evidently a fact-finding science that confines itself to the local level. Nevertheless, it is not concerned with the unique as such; although each locality on the earth's surface is climatologically unique from every other locality, the aim of microclimatology is the establishment of typical or characteristic climatic values for each locality. Yet, one is left with a problem respecting whether or not the distinction as drawn provides any basis for recognizing the two as distinct sciences, even if there is a sense in which microclimatology can be seen as a sort of concrete form of micrometeorology. The difficulty arises from the fact that the laws which micrometeorology seeks to establish do not spring from nothing, but rather are the result of the correlation of research carried on at a number of localities on the earth's surface. The typical or characteristic climatic values that microclimatology seeks to establish, therefore, can best be regarded as a necessary and logical step towards the establishment of laws, in that they provide the basic data needed for the establishment of the broader generalizations. If this is the case, then the setting up of two distinct sciences both dealing with the same data or subject matter, although differing in aim and level of generality, would appear to be a highly arbitrary matter without logical justification.

However, there is another sense in which a distinction among sciences in terms of "levels of generality" is sometimes made. For instance, one of the ultimate aims of the "unity of science" movement of the logical empiricists was the establishment of a few laws of such generality that the more specific laws of the biological and social sciences, as well as of the

---

[31]Gustav Hofmann, "Hints on Measurement Techniques Used in Microclimatologic and Micrometeorologic Investigations," in Rudolf Geiger, The Climate Near the Ground, pp. 520-522.

physical sciences, could be deduced from them.[32] The lower-order sciences of the hierarchy are evidently also law-finding sciences, but despite the hierarchical structure of the laws of the "unity of science," the end result is not simply one great science. Rather, individual sciences, and classes of closely related sciences, would continue to be recognized, presumably on the basis that the subject matter studied by, for instance, the biological and the social sciences is essentially distinct. One might also possibly want to maintain a distinction between levels within the hierarchy on the grounds that, for instance, the generalizations obtainable within the social sciences are less universal than those obtainable within the biological sciences, whereas in turn the generalizations obtainable within the biological sciences are less universal than those obtainable within the physical sciences. But this is a difficult notion to maintain, since it cannot be assumed that all the work done within a single science goes on at the same level of generality, or even that it aims at achieving generalizations that fall within a certain level of a conceivable hierarchy of the laws of science. For instance, although it might be maintained that the possible generalizations that could be achieved within the biological sciences are necessarily broader in terms of empirical coverage than those obtainable within the social sciences can ever be, on the grounds that man is but one biological species among many, having, moreover, a comparatively short history, much of biological research does not even aim at the establishment of generalizations of broad empirical coverage. This is especially true of the welter of "natural history" studies in biology.

Insofar as Kant viewed geography as a science of spatial relations, he regarded it, at least in part, as a law-finding science, and as a lower level form of physics. If we extend Kant's position to make allowance for the growth of science since his day, then geography in his view could conceivably be regarded as a lower level form of any science that studies phenomena occurring on the earth's surface. Hence, Kant's concept of geography as a science of spatial relations would come closest to corresponding to the view now under dis-

[32]For a succinct statement of the essentials of this position, see Rudolf Carnap, "Scientific Empiricism; Unity of Science Movement," in The Dictionary of Philosophy, ed. by Dagobert D. Runes, pp. 285-286.

cussion, that geography is a generalizing though lower-order or lower level form of science. Kant's distinction between higher level and lower level forms of generalizing sciences rests upon his distinction between theoretical and empirical sciences. The distinction, however, in the form in which Kant draws it, would probably not be acceptable to the majority of contemporary philosophers of science. The primary distinction drawn today is between rational or formal sciences and empirical sciences. The former include mathematics and logic; the latter include any science that deals with matters of fact. Hence, even "advanced" theoretical physics is an empirical science since at some point it deals with matters of fact.[33] Kant's distinction, however, does not involve a debate over the status of synthetic a priori propositions, since the laws of geography as well as those of physics must be exhibited in their conformity to synthetic a priori laws of the understanding, except in the case of the employment of mathematics. Since geography is a "description of nature" it cannot be a "system of nature" in the sense in which physics is, and hence it cannot employ mathematical propositions in the thoroughgoing manner in which the latter science can employ them. However, the Kantian thesis that the propositions of mathematics are synthetic and a priori is hardly acceptable to the vast majority of contemporary philosophers of mathematics and philosophers of science. On these grounds alone, his view that universal theoretical sciences, which can express their propositions completely in mathematical form, give us apodictic certainty, whereas particular empirical sciences can give us only degrees of hypothetical and probable knowledge, is highly questionable.

Nevertheless, if we set aside the question of the scientific and logical status of synthetic a priori propositions, there is yet a sense in which Kant's distinction between "universal" theoretical sciences and empirical sciences contains an element of truth. We ordinarily think of physics as being concerned primarily with the formulation of generalizations that are universal and abstract, that apply independently of any specific and particular situation, yet which can be applied, through deductive procedures, to given specific and particular situations that can be subsumed under, or exhibited as instances of, the general formulations. On the other hand, the low formula-

---

[33]For a succinct statement of this position, see Arthur Pap, _An Introduction to the Philosophy of Science_, p. 139.

tions of geography and the social sciences, if such exist, are "regional," and are, by their very nature, limited to the explication of relations among phenomena that pertain only to specific spatio-temporal situations, i.e., that are destined to persist over only limited periods of time and that pertain to only limited portions of the earth's surface. However, the same "regional" limitations apply to the general statements that can be formulated by the physical earth sciences. For instance, the geomorphologist who attempts to explain how "V"-shaped valleys are transformed into "U"-shaped valleys is dealing with phenomena found only in mountainous areas that have been subjected to glaciation, and that will persist on the earth's surface for only as long as the glaciated mountain systems themselves persist. The geomorphologist, of course, is applying a number of physical laws to arrive at his explanation. Such laws would include the law of gravitation, and physical generalizations pertaining to the erosive powers of glacier ice, the tendency of a plastic solid like glacier ice to erode to a hemispherical shape that will most readily accommodate the greatest volume of material, the erodability of various types of rock surfaces, etc. Yet, despite the fact that the geomorphologist is constantly applying physical generalizations in his research, we do not ordinarily think of the study of the landforms of the earth's surface as being a branch of physics. Geomorphology, at least on the basis of the example cited, would appear to be more a law-applying than a law-finding science, since we cannot, in any strict sense, talk of a single law that explains the transformation of "V"-shaped valleys into "U"-shaped valleys, despite the fact that the geomorphologist may arrive at some generalizations respecting the matter. Rather, the explanation proceeds on the basis of the application of a variety of physical generalizations to a specific type of landform. And insofar as the geomorphologist is concerned with how one glaciated valley may differ from another, or with what is peculiar to a given glaciated valley, he is concerned to some degree with the variable ways in which the physical generalizations may apply and modify one another under actual specific conditions existing on the earth's surface. Thus there is a sense in which geomorphology may be thought of as a lower-order form of physics.

However, the possibility of regarding geography as a lower-order form of the various biological and social sciences presents greater difficulties, since the laws of biology and the social sciences are "regional" by their very nature. In the case of considering geomorphology as a lower-order form of physics, at least some distinction can be made between the universal generalizations of physics and the regional generalizations of geomorphology. It is difficult, however, to attach precise meaning to the idea that the generalizations that may possibly be achieved in biogeography, economic geography, political geography, etc. are lower-order biological, economic, and political scientific generalizations. Geographical generalizations are often not deducible from broader biological and social scientific generalizations, nor are they arrived at through the application of these generalizations to specific geographical situations. For instance, generalizations concerning the distribution of biological organisms over the surface of the earth cannot be arrived at in such a manner and hence should be regarded as lower level rather than as lower-order statements. As indicated in the previous chapter, however, such studies can perhaps best be regarded as parts or adjuncts of biology, economics, etc. Biogeography consists principally of the study of (1) the distribution of biological organisms over the surface of the earth, (2) mutual spatial relations among various biological species, and (3) the relations of biological organisms to their physical environments. (1) and (2) are clearly branches of biology, since their central focus is biological, since they presuppose considerable biological knowledge, and since the mere addition of space in such studies adds nothing specific by way of non-biological content; they are not the concern of an entirely separate science, geography, despite the clear recognition that they are geographical sciences in the broad sense of that term. The status of (3) is debatable, since in this type of study the biologist is involving himself to a considerable degree in the study of non-biological phenomena, such as climate, soils, rocks, etc. There is a peculiarly "synoptic" quality to this type of "ecological" study, and since geography often exhibits this "synoptic" quality, a case can be made for regarding such studies as more geographical than biological, or at least as much geographical as biological. Nevertheless, although the logical status of such studies is far from clear, their central focus remains biological organisms, and the biologist at least has the distinct advantage of beginning his "ecological" studies from a solid basis of biological knowledge. Similarly, studies of the distribution of economic goods over the surface of the earth, or of spatial interaction among types of economic enterprises or functions would appear to be logically a part of the content of the science of economics. Geography, insofar as it carries out such studies, is really performing specific tasks that come within the framework of the science of economics.

d. Geography as the Science that Studies "Things in Reality" Spatially

Hartshorne[34] has referred to geography and history as "naive sciences" that "attempt to integrate all kinds of phenomena in space or time, " and that look "at things as they are actually arranged and related, " rather than from the more "sophisticated, " "artificial, " abstractive, and limited point of view of more specialized sciences, "which take phenomena of particular kinds out of their real settings. " Although he has in mind in this context a fundamentally different concept of the nature of geography, that of "areal differentiation and areal integration, " and although he tends to regard geography as being primarily concerned with the study of phenomena that are unique or non-repeatable, he rejects any absolute distinction between nomothetic and idiographic sciences, and thereby does not on principle rule out the possibility that geography may be a generalizing or law-finding science (pp. 378-379). Thus, the possibility is raised that geography may be conceived as a law-finding science that differs from other such sciences on the grounds that it deals with spatial relations among "things in reality. " However, the idea that geography studies "things in reality" raises some major difficulties. For instance, the idea is to some extent belied by the prevalent contemporary tendency to divide geography into branches according to widely recognized scientific categories, e.g., economic, political, and biological. On the basis of this division, geography studies the same objects as do other sciences, and not "things in reality" that are in some sense different from the things studied by other disciplines. In addition, the idea that geography studies "things in reality" implies that other sciences differ from geography in that they abstract from "reality. " But obviously, the geographer does not and cannot study everything,

[34]The Nature of Geography, p. 373.

and in this sense he must also abstract from "reality." And in fact it can be argued that since the perspective of geography is often broader than that of more specialized sciences, then geography will inevitably abstract from "reality" to an even greater extent than do other surface-of-the-earth sciences. Since geography is abstractive, and since, on this view, it is a generalizing science that "attempt[s] to integrate all kinds of phenomena in space," this concept of geography appears to be indistinguishable from the earlier concept that geography is a super-science of spatial relations. In addition, it is difficult to see why this type of study is any more "real" than the type of study that limits itself to matters purely economic, political, or social. "Reality" in this context appears to imply a condition of "organic totality."

Nevertheless, there is perhaps a sense in which geographers can be thought of as studying "things in reality," since they often study spatial relations among things that do not readily fit under the sets of relations that other sciences study. Thus a geographer may be interested in studying the relations between residential districts, major thoroughfares, traffic flow, and business districts within urban centres. We do not ordinarily think of houses, residential districts, streets, traffic flow, and business districts as being objects of study for any particular science. However, a drawback to this interpretation is that geography then becomes a "bits and pieces" subject, destined to be limited to the study of spatial relations among phenomena that do not come readily within the jurisdiction of other sciences. Furthermore, in this case, there is nothing restricting the emergence of a science of "urbanology" that studies urban centres in all their ramifications. And, in fact, the growing number of institutes of urban research throughout the world might indicate that a science of "urbanology" is emerging. Few if any people, however, are trained in "urbanology" as such; rather it appears to be a synoptic discipline that attracts town planners, architects, civil engineers, sociologists, and economists as well as geographers. If a science of "urbanology" does arise that has urban centres as its object of study, then urban geography will be relegated to the carrying out of certain specific tasks within the framework of such a science. Finally, it is difficult to see how the type of study that the urban geographer may engage in, even if in a sense he is concerned with the study of objects that differ from those that ordinarily concern other scientists, is any more concerned with the "real" than are economic, sociological, or political studies.

(e) Any consideration of the proposition that geography is not a generalizing or law-finding science of spatial relations leads us into a different interpretation of the concept "spatial relations," and so to a different conception of what geography is fundamentally about.

3. AREAL DIFFERENTIATION
    AND AREAL INTEGRATION

While superficially this third concept of the nature of geography appears to relate closely to the second, since it also rests upon a distinction between the classes "systematic" or "subject matter," spatial, and temporal sciences, its origin and basis is quite different. The basis for the concept has often been expressed by geographers. The following passage from Hartshorne will suffice:

It should further understanding of the concept to remind ourselves, in elementary terms, of the basic reasons for the existence of a field called geography. If the combinations of factors found in any area of the earth—the particular conditions of climate, landforms, soils, population, crops, farms, cities, and so forth—were found to be almost the same in every other area, geography would be reduced to the problem of determining the interrelationships among these diverse factors which produced the total complex repeated without variation over the world. Under these conditions, such a subject, if it existed at all as a separate discipline, would no doubt have developed relatively late and as an overall integrating science of little popular interest.

Very early in human development, however, man discovered that his world varied greatly from place to place. It was to satisfy man's curiosity concerning such differences that geography developed as a subject of popular interest. From earliest times travelers returning from "foreign" parts were expected to tell the stay-at-homes what things and people were like in the places they had seen, whether in adjoining, but relatively inaccessible districts or in more remote parts.

This universal curiosity of man about the world beyond his immediate horizon, a world known to differ in varying degrees from the home area, is the foundation of all geography. [35]

[35]Hartshorne, Perspective on the Nature of Geography, p. 15.

Although as Hartshorne himself goes on to point out, this orientation in geography does not necessarily preclude interest in similarities between things from place to place, or area to area, over the earth's surface, what strikes the geographer most is the fact that the surface of the earth is not uniform, but rather differs from area to area in terms of any number of factors as well as in terms of how the various factors fit together. However, the geographer, or anyone else for that matter, is not interested in every minute point on the earth's surface, and in how every minute point differs from every other minute point. Rather, he is interested in areas of the earth's surface that appear to be more or less homogeneous with respect to one or more factors, yet which can be differentiated from other areas where the combinations of factors fit together differently. The term that is most commonly employed by geographers to describe these areas is "region." Hence, if geography is conceived as the science of areal differentiation and areal integration, "region" is the basic concept. This fundamental orientation towards regional analysis tends to differentiate this concept of the nature of geography rather sharply from the second one, and in fact some holders of the concept of geography as the science of spatial relations and spatial distributions regard their orientation as "freeing the field from a past view of the world as a mosaic of regions."[36]

It is not necessary, however, to offer a detailed analysis of the concept "region," and of the variety of ways in which it has been employed by geographers. This has been done often enough by geographers themselves, and in fact they have paid far more attention to the concept than have the members of any other discipline. A considerable body of geographical literature has grown up around this concept.[37] In addition, the concept is not of central importance for Kant. Although the work of Gatterer and others on devising "natural" regions was proceeding during the last quarter of the eighteenth century,[38] Kant gives no indication that he was ever influenced by it. Because of his acquaintance with a wide range of geographical

[36]Report of the Ad Hoc Committee on Geography, The Science of Geography, p. 12.

[37]The most comprehensive recent treatment of the concept "region" is to be found in Roger Minshull, Regional Geography: Theory and Practice.

[38]For an account of Gatterer and "natural" regions, see Hartshorne, The Nature of Geography, pp. 37-38, 250.

literature, it is difficult to imagine that he was not aware of this work. Probably it did not suit his purpose of giving a popular introductory course in geography. Apart from the fundamental physical division of the earth into continents and oceans, undoubtedly the way in which most people regard the world as divided into regions is in terms of political units. Kant's so-called regional geography consists of little more than descriptions of various countries. "Natural" regions, as devised by Gatterer, was probably a somewhat esoteric topic at the time.

Perhaps the most widely quoted recent passage on the nature of a region is the following:

. . . the region [is] a device for selecting and studying areal groupings of the complex phenomena found on the earth. Any segment or portion of the earth surface is a region if it is homogeneous in terms of such an areal grouping. Its homogeneity is determined by criteria formulated for the purpose of sorting from the whole range of earth phenomena the items required to express or illuminate a particular grouping, areally cohesive. So defined, a region is not an object, either self-determined or nature-given. It is an intellectual concept, an entity for the purposes of thought, created by the selection of certain features that are relevant to an areal interest or problem and by the disregard of all features that are considered to be irrelevant.[39]

Kant would almost certainly have agreed that a region, or portion of the earth's surface, is not "nature-given," that is, that it is not something which can be given to an observer in sense experience; it is not an object of experience. To put the matter in Kantian language, a region is an Idea, just as the world itself is an Idea.

However, it would appear that there are several senses in which one can meaningfully talk about "natural" regions. For instance, in a purely physiographic sense, the surface of the earth can be divided up quite readily into a variety of clearly recognizable landform types—mountains, hills, plains, etc.

[39]Derwent Whittlesey, "The Regional Concept and the Regional Method, " in American Geography: Inventory and Prospect, ed. by Preston E. James and Clarence F. Jones, p. 30. However, an alternative basic concept of a region, which still has a number of adherents, is supplied by Paul Vidal de la Blache, who disputes the idea that "homogeneity" is an essential requirement of a region. Vidal maintains that a region may be heterogeneous with respect to its contents, yet may exhibit as a whole a unity within diversity, and may reveal a certain "character" or "personality" that is peculiarly its own. See, for instance, his Principles of Human Geography, pp. 17-18.

Again, if a person were to travel across Canada, he could not help but be aware of striking areal differences from one part of the country to another. In passing from the Canadian Shield onto the prairies, one is acutely aware, within a relatively short distance, not only of a difference in the appearance of the landscape, but also of differences in the way in which man lives and organizes the space at his disposal. One might not arrive at the "Idea" of a region, but one could not help but be immediately conscious of profound differences between areas or portions of the earth's surface.

Ordinarily, the regional geographer aims at far more than the delineation and understanding of regions that are based on a single criterion. Since such studies as regional economics, regional sociology, etc. can be recognized, that is, the study of economic and social problems within given regional contexts, then regional geography must aim at more, if geography is to maintain itself as a separate discipline. The tendency then, as Hartshorne argues, is for the regional geographer to aim ideally at "totality" or "total integration of data." Obviously, however, the regional geographer cannot include everything in his study, anymore than can the historian who attempts to present a comprehensive picture of life during a certain historical epoch, period, or century. Depending in part upon the size of the region being studied, the regional geographer often aims at providing a comprehensive understanding of the spatial interrelations between large scale types of phenomena— major landforms, climates, biotic types, farming types, cities, etc.

It has often been pointed out by regional geographers that the type of study in which they engage is more closely related to the work of the historian who provides an overview of an historical period than it is to work in any other discipline. The history of an epoch or period attempts to provide an overview, a more general account than would be given by any one of the more specialized sciences. It attempts to provide a "feeling" for what life must have been like during that epoch or period. No very clear distinction, however, can be drawn between the two types of study on the grounds that the one deals exclusively with spatial matters whereas the other deals exclusively with temporal matters. For instance, the historian who gives us an overview of the "Golden Age of Greece" is not only concerned with a series of events which lasted through a definite period of time, but which also extended over a definite portion of the surface of the earth. Nevertheless, he will be much less concerned than would the geographer with the actual spatial association among things. On the other hand, unless he is doing historical regional geography (i.e., studying the spatial associations among things in some area at some definite point of time in the past), the regional geographer will ordinarily be interested in the present, although he must assume at least a specious present in order to discuss influences over distance that inevitably require time in which to happen. However, the regional geographer cannot assume the time dimensions that the historian does. In other words, he is not primarily concerned with how any particular series of occurrences has changed over time; this kind of study is presumably the concern of either the historian or of the historical dimension of some other science. In addition, it makes little sense to talk about the historical development over time of a region as a whole, especially if a region is not a thing or object, but rather an "idea," and at best a heuristic device. Similarly, it makes little sense to talk about the development of an epoch as a whole, although the historian can talk about developments and changes that occur within the given epoch, since the epoch is only the "idea," the label the historian applies to designate the degree of unity he hopes is present so that the material he presents may be seen and understood as a more or less coherent whole.

However, this tendency to treat regions and epochs as "wholes" presents difficulties. It can lead to a disregarding of the fact that a region is often only an "idea" or higher-order concept, and to a tendency to treat regions as things or as lower-order concepts. For instance, a recurrent theme among geographers throughout the last 150 years is that regions, on a close analogy with biological organisms, are "living organisms" or "organic wholes."[40] If regions are regarded as "organic wholes," then only a short step is required to reify the concept and to start talking about regional interaction, or about regions or cities interacting with one another. Things within one region or city may interact with things from another region or

[40]For a critical review of the history of this concept of a region, see Hartshorne, The Nature of Geography, pp. 256-260.

207

city, but not the regions and cities themselves as wholes. Are there any alternatives to regarding regions as "ideas" or "wholes"? As indicated earlier, there is evidently a sense in which we can talk about "natural" regions; i.e., we can be perceptually aware of differences between portions of the earth's surface, and not merely in a purely physiographic sense, but also in that we can be immediately aware of spatial organizational differences. Nevertheless, it is still not the region we are perceiving, but things within the region. We cannot perceive a "natural" region in the sense in which we can perceive trees, lamp-posts, and cats. Trees, lamp-posts, and cats can be perceptually isolated from other things, whereas regions cannot. Furthermore, some regions cannot be regarded as "natural" in any sense; some would appear to be almost purely "ideas." For instance, the United States could be divided up into regions on the basis of the circulation of major daily newspapers, and the boundaries between such regions could be delineated with a fair degree of statistical accuracy, yet in no sense could we be perceptually aware of such regions. Even if some regions can be regarded as "natural" in a sense, whereas others appear to be more purely "ideas," is there any middle ground between the extremes of regions as conceptual "wholes" and regions as "natural" entities? A possible alternative is that regions are simply portions of the earth's surface that may be delineated and separated from one another either statistically or intuitively. This expedient can avoid the difficulties involved in regarding regions as "wholes" or even as "organic wholes" that interact with one another and that necessarily exhibit a high degree of internal cohesiveness. It can also avoid the difficulties involved in attempting to maintain that regions are "things" or "natural entities." The alternative concept still allows the geographer to study the spatial associations among things within his chosen region. Furthermore, the degree of internal cohesiveness within the region is left as an open question, as something to be explored, and is not prejudged as is the tendency if one regards regions initially as "organic wholes." Finally, this alternative concept allows the geographer greater freedom in exploring the temporal aspects of his chosen region, since, if regions are regarded as "ideas" or "intellectual concepts," one cannot meaningfully talk about the development over time of what is only an "idea." However, if a temporal dimension is included in regional studies, then the focus of the geographer must be different from that of the historian so that the one type of study can be distinguished from the other.

The historian of an epoch or period will ordinarily devote much space to the mental achievements of man, such as the philosophy, science, art, literature, music, etc., of a period, yet he ordinarily has little to say, except possibly by way of an introductory chapter, on the landscape features, climate, biotic resources, etc. of the portion of the earth's surface encompassed by the historical period. On the other hand, the regional geographer, although having virtually nothing to say on the mental achievements occurring within his region, or on purely social matters, will have much to say on landforms, climates, biotic types, natural resources, industries, cities, transportation routes, etc. — in other words on matters that can be spatially located and related. However, all this is but what one would normally expect, given the orientation of the historian towards human history and towards predominantly temporal events over spatial events. For instance, the great mental achievements of mankind possess a peculiarly temporal quality. To know that Plato spent most of his life in Athens and visited Sicily in old age, or even to know the typical places of residence and the itineraries of the philosophers with whom he associated, does not provide much help in understanding his philosophy. On the other hand, the regional geographer is not really concerned with men or human actions as such. He is much more concerned with areas or portions of the earth's surface, and things contained therein, such as landforms, biotic types, cities, major transportation routes, farming types, etc. —things with some degree of concreteness and permanence that can be shown to stand in mutual relations of spatial association with one another.

One might inquire as to the scientific status of regional geography and epochal history, or indeed one might question the application of the label "science" to either of them. Given the scope and level of generality of each, it can hardly be maintained that they are exact sciences in the sense in which we accord such status to sciences that deal with a much more specific and limited subject matter, and which can verify some of their results experimentally. Epochal history and regional

geography, if exact sciences, could only qualify for super-
science status, which of course neither of them warrants. As
indicated earlier, Hartshorne[41] has described history and
geography as "naive sciences" that "attempt to integrate all
kinds of phenomena in space or time," and that look "at things
as they are actually arranged and related." Or, to put the
matter in Kantian language, history and geography, in one of
their interpretations, are "rough" disciplines, whose aim and
perspective is different from that of the more "exact" sciences.
Neither particularly aims at the establishment of general laws
as do a number of the more "exact" sciences, but rather at
providing an overview in the form of some degree of coherence
of either knowledge or experience at the empirical level, which
no particular specialized science can achieve.

4. PLACE AND LOCATION

a. Place
The idea that the central concern of geography is the study of
"places" has been enshrined in Vidal de la Blache's somewhat
elliptical remark that "La géographie est la science des lieux,
non des hommes."[42] Nevertheless, the concept "place" has
been used by geographers in at least three different senses,
or at three different levels. In the limited and perhaps proper
sense, the concept refers to a specific and particular part of
space, and to what may occupy that space, as when we think
of our place of residence as being a particular building, or as
when we talk of a place of worship or a place of amusement.
In a broader sense, the concept is sometimes used to refer to
a more extended portion of the earth's surface, such as a city,
a province, or even a country. In this sense, the concept
"place" appears to be indistinguishable from the concept "region." In an even broader sense, the concept has occasionally
been used to refer to the whole surface of the earth. For instance, Vidal de la Blache[43] thinks of the surface of the earth

---

[41] The Nature of Geography, p. 373.

[42] Vidal de la Blache, "Des caractères distinctifs de la géographie," Annales de
Géographie, XXII (1913), 289-299; quote from p. 298. Vidal, however, includes man
as a part of place.

[43] Principles of Human Geography, p. 6.

as forming a "terrestrial unity, " as being "the stage upon which man's activities take place, " and hence as the "place" of man.

Only in the first sense, or at the first level, would the concept "place" appear to offer even the possibility of a distinctive conception of what geography is about. But even in this case, since geography is concerned with spatial relations, and since places are not points but exhibit some degree of spatial extension, it can be argued that they are actually very small regions of the earth's surface, and hence that the concept "place" employed even in its restricted sense is not capable of being clearly distinguished from the concept "region, " which in its widest interpretation encompasses any portion of the earth's surface that possesses some degree of spatial extension.

Nevertheless, there would appear to be at least one sense in which the concept "place" is perhaps distinguishable from the concept "region. " The former concept appears to possess a perceptual unity that the latter concept lacks. For instance, if I ask myself the question, "What is my place ?" I might, without much reflection, reply that it is the city of Toronto. However, many parts of Toronto I have never visited, and hence have never seen, despite the fact that I have lived in the city for more than half my life; these unseen parts of Toronto are not "my place. " Evidently, I mean something much more specific by "my place. " Yet, "my place" is more than the room in which I habitually work, more than the apartment I rent, and even more than the building in which my apartment is located. As I look out my front window, or stand outside the apartment building, I perceive a number of things spatially associated with one another. It is, I think, this familiar association of things, together with the apartment building in which I live, which I can take in at a glance, that constitutes "my place. " However, because of human mobility, rapid urban structural changes, and the prevalent tendency of urban dwellers to divide their lives between a place of residence and a place of employment, it is debatable whether many people living in large urban centres have a very deep-rooted sense of place. A rural setting provides a better illustration of the concept "my place. " For instance, my uncle has farmed the same land in Southern Ontario for most of his life. From his house, the interlobate moraine of southcentral Ontario rises gently to its crest, a

211

few hundred yards to the north. A couple of miles to the south, across a broad valley, is a recessional moraine. A kame stands out rather prominently to the southwest. This landscape is dotted with trees, fields, fences, and various farm buildings. One can take it all in, with a glance or two, in a matter of seconds. To my uncle, this is his "place." It is more than just the farmhouse, more than any one feature of the landscape. Although a few years ago the original farmhouse burned down and the orchard was cut down, his place remains even if altered somewhat, for a "place" is a unique spatial association of things that can be taken in at a glance, that possesses a certain perceptual unity. However, if the buildings were destroyed, the trees cut down, the kame removed, and the moraine levelled to fill in the valley, in short, if the whole perceptual landscape were obliterated, and even if my uncle were to be relocated in a neighbouring valley, he would feel very strongly that his "place" was gone.

Although most geographers would probably not hold the concept "place" in quite the specific sense in which I have attempted to define it, but would prefer to use it more broadly, what the concept does raise is the view that geography is concerned with the study of unique cases, for places are unique portions of the earth's surface. However, regions have also often been regarded by geographers as being unique entities, as exhibiting the quality of uniqueness. Yet, regardless of whether the concept "place" is employed in its very restricted sense, or the concept "region" is employed to designate a more extended portion of the earth's surface, what the issue of uniqueness raises is the possibility of setting geography off rather sharply from other surface-of-the-earth sciences. For, insofar as geography is regarded as studying particulars, since particulars often lead to the establishment of broader generalizations, and since particulars are often subsumable under broader generalizations or can be exhibited as instances of such generalizations, then geography becomes merely a lower-order form of a variety of other sciences. In addition, as indicated in an earlier context, the difficulties of establishing geography as a generalizing or law-finding science of spatial relations are great. However, insofar as geography is concerned with "places" or "regions" as unique portions of the earth's surface it is concerned with entities that are not subsumable, even as particular cases, under broader generalizations, for what we mean essentially by uniqueness is something that is non-repeatable. Hence, the geographer of particular places and regions is concerned primarily with providing a description of the unique configuration or spatial association of things as they are found in a specific portion of the earth's surface. Insofar as such a discipline can be explanatory, it would appear to employ an historical form of explanation, i.e., to account for the present unique configuration of things on the basis of how that configuration has come to be as it now is. Thus, the geography of "places" and "regions" bears a close relation to history in the specific sense, i.e., a history that aims at providing an understanding of a unique historical event, whether a significant particular action of some historical personage, or an "histor-ical event" in the broader sense of that term, such as the French Revolution. Such a history, although it might employ generalizations from other sciences as aids to understanding, and thus to a degree be law-applying, does not ordinarily aim at the establishment of generalizations or even necessarily at the explication of particulars that might be exhibited as in-stances of broader generalizations. For instance, the political scientist or the sociologist would be interested in studying revolutions as a type of political or social occurrence, and would be interested in establishing generalizations respecting that type, whereas the historian, as historian, would be pri-marily interested in understanding an instance of the type, such as the French Revolution. And much of the information he could collect respecting that revolution would not pertain, even as particular instances, to revolutions in general, but would remain unique to the French Revolution.

With respect to this matter, Kant indicates, in the introduc-tion to the <u>Physische Geographie</u>, that the division of geography which is called "topography" is concerned with "the description of a particular place on the earth." Furthermore, in the same context, Kant points out that in geography "things are observed according to the places which they occupy on the earth." Al-though geography is evidently concerned with places, as an initial prerequisite for the subject, it is not concerned with what is unique, since, for Kant, there can be no science of the unique.

b. Location

To locate something is to pin-point it exactly. Hence, the concept "location" has even more specific reference than has the concept "place" as defined above, for a place is made up of a number of things that can be specifically located, and of the spatial relations among those things. In this quite specific sense of the concept, there is nothing peculiar to geography about location. If we are to talk about things in an empirically meaningful way, then we must be able to locate those things. Hence, location is the concern of every science. The determination of the location of any specific object or set of objects, such as an industry, a type of social institution, a species of animal, etc., is logically a part of the content of the particular science which deals with that object or set of objects. In short, no separate science of location is possible, except possibly in the case of mathematics (e.g., geodesy), and geography cannot be defined as the science of location.

Nevertheless, geographers have often used the concept "location" more broadly, and have attempted to define their discipline as a science of location. For instance, Lukermann has recently remarked that

> the first insight, the first conceptual theme of geography, is location. . . .
>
> Geographers have given variant meanings to location in the specific context of their studies, but in general these have been given in answer to one question, that of place. . . . the place of something has been described in terms of the internal arrangement of features (site) and of external connectivity and environs (situation). Separately or together as definitions of place, site and situation are locating in relation to some other place or thing. To locate is to relate; and a relationship is a conceptual system of thought. This meaning of location to the geographer—the relative distributional condition of the area under study—is the first conceptual criterion of doing geography. If the geographer fails to conceive of his research as involving this step it seems difficult to accept such study as either explanatory or descriptive in any geographical sense.[44]

The concept "location," as employed by Lukermann, is indistinguishable from concepts such as "place," "region," or "area," since places, regions, and areas are made up of things that can be spatially related to one another. The ability to locate these things can be taken for granted as an initial prerequisite to

---

[44]Lukermann, "Geography as a Formal Intellectual Discipline and the Way in which It Contributes to Human Knowledge," The Canadian Geographer, VIII, No. 4 (1964), 167-172; quote from p. 169.

studying places, regions, and areas. Furthermore, the concept "location" as employed in this context implies the study of spatial relations, whether in the sense of the distribution of things, or in the sense of spatial interaction among things.[45] Hence, the concept "location" as employed in this extended sense appears to be indistinguishable from other concepts of the nature of geography that have been discussed previously.

One sometimes encounters the phrase "theory of location," as, for instance, when Alfred Weber talks of a "theory of the location of industry." "Theory of location" in this context appears to involve: (1) an account of the spatial patterns that can be discovered among a variety of industries, or of whatever other objects may be patterned spatially; and (2) the prediction of the location of specific instances of whatever is studied, insofar as this is possible in any precise sense. "Theory of location" is thus logically a part of the content of whatever science studies a particular set of objects. Many economists, for example, at least from the time of von Thünen on, have been vitally concerned with the "theory of location" of a variety of economic matters. And insofar as one can even talk about geography as concerned with a general theory of location, i.e., a single theory that somehow accounts for the location of everything on the earth's surface, one would again be involved in the notion of geography as a sort of super-science.

Furthermore, as in the case of the question of place, the question of location may involve the researcher in the issue of uniqueness. For instance, one may ask the questions: "Why did Henry Ford locate his automobile plant in Detroit?" or "Why did B. F. Goodrich locate his rubber plant in Akron, Ohio?" Neither specific location is in any strict sense subsumable under general economic principles of location, since either plant could have been located elsewhere within the industrial heartland of the United States, and could have been successful economically. Rather, one needs to enquire into the whole complex of circumstances that led Henry Ford to locate in Detroit and B. F.

[45]For instance, Bunge, Theoretical Geography, p. 195, defines geography most generally as "the science of locations," yet his whole treatment of the discipline indicates quite clearly that he regards it as the science of "spatial relations." Again, Peter Haggett, Locational Analysis in Human Geography, pp. 12-13, although evidently a member of the "locational school" in geography, regards location studies as basically "spatial," and moreover, points out that such studies are to some degree the concern of all sciences and not the exclusive prerogative of geographers.

Goodrich to locate in Akron, Ohio. In these cases, one would appear to be involved in socio-psychological and historical matters, as much as one is involved in specific geographical matters that are not subsumable under general statements.

5. SYNOPTIC

Geographers have often vaguely referred to their discipline as the "correlative science," the "science of interrelationships," or as a "liaison discipline."[46] However, geography may be regarded as a synoptic discipline in two distinct senses: (a) research and (b) academic.

a. Research
In a research sense, the concept "synoptic" does not add anything distinctive by way of a fundamental concept of the nature of geography. For instance, environmental studies are certainly synoptic in the sense that they attempt to relate man to his physical environment, and thereby attempt to correlate data derived from the physical and social sciences. Furthermore, "ecological" studies are clearly synoptic, in that they attempt to correlate data that are arrived at by a number of specific and specialized disciplines. Because of the often peculiarly "synoptic" quality of their discipline, there is no inherent reason why geographers cannot engage in this type of study. However, the geographer often suffers the major disadvantage of a lack of grounding in at least one specific subject-matter science, and hence his synoptic studies in the "ecological" sense usually tend to be rather superficial. Furthermore, regional studies are certainly synoptic in a loose sense, in that they often provide us with an overview of the spatial association among diverse things contained within a portion of the earth's surface.

---

[46]For instance, see Griffith Taylor, "Geography the Correlative Science," Canadian Journal of Economics and Political Science, I, No. 4 (1935), 535-550; Our Evolving Civilization, p. 4; and "Introduction: The Scope of the Volume," Geography in the Twentieth Century, pp. 3-27.

b. Academic

Strahler has argued that there is no longer such a discipline as physical geography, in the sense that no one any longer does research in physical geography:

> What is physical geography? As a first step in understanding this term we might well expand it to read "the physical basis of geography," for physical geography is simply the study and unification of a number of earth sciences which give us a general insight into the nature of man's environment. Not in itself a distinct branch of science, physical geography is a body of basic principles of earth science selected with a view to including primarily the environmental influences that vary from place to place over the earth's surface.[47]

Although the above quotation indicates that Strahler accepts "environmentalism" as the fundamental concept of geography, it is clear that he regards physical geography as a synoptic discipline in an academic and not in a research sense. In fact, he goes on (pp. 1-2) to point out that physical geography is a "unification" of elements of geodesy, astronomy, cartography, meteorology, climatology, pedology, plant geography, physical oceanography, geomorphology, geology, and hydrology—all of which he regards as distinct scientific disciplines—that aims at achieving an overview of the surface of the earth, the environment of man.

If physical geography has disappeared in a research sense, and remains only as an academic discipline, one is left to wonder respecting the status of human geography. Does it survive simply because of the underdeveloped state of the social sciences?

There can be little question about the fact that geography played an extremely important role for Kant as a synoptic discipline in the academic sense, especially in the later years of his teaching at Königsberg. It provided a framework, an integrated, even if somewhat overgeneralized, overview of the surface of the earth, the environment of man. If such a discipline was necessary in Kant's day, it is even more necessary today, given the extreme compartmentalization of knowledge characteristic of the twentieth century.

[47]Arthur N. Strahler, Physical Geography, Introduction, p. 1.

6. APPLIED

There has been a recent tendency among some geographers to regard their discipline as an applied science, in the sense that it deals with practical problem-solving in connection with resource management, water utilization, optimum agricultural land-use, etc. This position is strengthened by the fact that some aspects of contemporary geographical research appear to amount to little more than the application of principles and generalizations from a number of other sciences to practical problem-solving situations. If geography is nothing more than an applied science, and possesses no general principles of its own, then it should be classified along with disciplines such as social work and engineering. Social work consists basically of the application of principles and generalizations drawn from psychology and sociology to the solving of practical social and social-psychological problems. Engineering consists basically of the application of principles and generalizations drawn from physics and chemistry to the production of any number of physical social amenities. However, if "environmentalism," "regionalism," "place," and "synoptic" in the academic sense are legitimate concepts that establish geography as an independent discipline, then geography obviously is not a purely applied discipline, but has some claim to independent status.

SUMMARY

The foregoing discussion leads to the conclusion that geography can establish its independence as a distinctive discipline in at least four senses: (1) "environmentalism" in the sense of the study of man in relation to his physical environment; the justification, however, appears to be pragmatic and not logical, since, although few other disciplines have claimed this kind of study, a social science could justify itself in doing so because biology has established a precedent; (2) "regional" studies—in this case the geographer is attempting to provide an overview of the spatial association among things as it occurs in any portion of the earth's surface and at any time; no other discipline attempts to do this; (3) the study of "place," in the sense of the unique perceptual integration of phenomena as it occurs within any limited portion of the earth's surface; and (4) "synoptic" in the academic sense—geography combines elements from the findings of a number of other disciplines to provide an integrated overview of the surface of the earth.

On the other hand, several concepts of the nature of geography appear to lack adequate logical justification: (1) geography as "ecology," in the sense of the study of the spatial association among diverse phenomena, suffers from the major weakness that geographers tend to lack adequate grounding in a specific subject-matter science; in this case, the lack of justification is pragmatic rather than logical; (2) geography as a generalizing science of spatial relations suffers from the fact that all sciences are concerned with spatial relations, and that such studies are logically a part of the content of any subject-matter science that deals with a distinctive set of objects; and (3) geography as a science of location suffers the similar disadvantage that all other sciences are inevitably concerned with the question of location. No separate and distinct empirical science of spatial relations or location is logically possible.

IX
The place of geography in a classification of the sciences

If it becomes desirable to organize any knowledge as science, it will be necessary first to determine accurately those peculiar features which no other science has in common with it, constituting its peculiarity; otherwise the boundaries of all sciences become confused, and none of them can be treated thoroughly according to its nature.

The peculiar characteristics of a science may consist of a simple differ- ence of object, or of the sources of knowledge, or of the kind of knowledge, or perhaps of all three conjointly. On these, therefore, depends the idea of a possible science and its territory.[1]

Kant's first criterion for distinguishing sciences, "a simple difference of object, " is undoubtedly the one most commonly employed. We customarily recognize zoology, botany, geology, etc. as differentiated on the basis of the kinds of objects they study. This criterion, however, presents major difficulties for finding a place for geography among the sciences, since it is generally agreed upon by geographers that their discipline has no objects of study that are peculiarly its own. Also, for Kant, as we have seen, geography cannot be differentiated

[1]Kant, Prolegomena to Any Future Metaphysics, sec. 1.

from other sciences on this basis, since by and large it studies the same objects as does physics. Kant's second criterion, "the sources of knowledge, " is the one that he himself employs to differentiate geography from physics. This criterion rests on his distinction between theoretical and empirical sciences, which was discussed to some extent in the previous chapter. In addition, this criterion may refer to his distinction between outer and inner sense, since some sciences, such as physics and geography, rest upon a combination of outer and inner sense, whereas other sciences, such as psychology and anthropology, rest upon inner sense alone.

Kant's third criterion, "the kind of knowledge, " remains extremely vague, and is not enlarged upon by him. In Kant's case, this criterion could conceivably refer to his implicit distinction between research and academic disciplines. In an important respect, geography and anthropology were, for Kant, academic disciplines, and hence could be differentiated from research sciences, or geography as a research science could be differentiated from geography as an academic discipline. Furthermore, the idea of "the kind of science" could conceivably encompass such criteria for differentiating sciences as: (a) a difference in methodology; (b) a fundamental difference in perspective; or (c) a difference between classes of sciences such as law-finding, "typological, " and fact-finding.

(a) Kant gives no indication that there is anything methodologically unique about geography, especially in a research sense, except for the fact that geography cannot employ mathematics in the thoroughgoing manner in which physics can, and hence must rely on the employment of hypotheses and analogies. But these procedures it holds in common with the other empirical sciences. It has been repeatedly denied by contemporary geographers that there is a distinctive "geographic method. " For instance, Carl Sauer[2] has laid to rest the myth of a unique "geographic method, " i.e., mapping "the distribution or variation of anything that is spatially localized or varying, " on the grounds that any number of other surface-of-the-earth sciences also employ maps as vital methodological tools. More recently, Lukermann has argued that

2"Folkways of Social Science, " in Leighly, ed., Land and Life, pp. 380-388; see especially p. 381.

221

as a science it [geography] has no particular method of investigation. It observes, measures, classifies as every other science; in other words, it employs the scientific method which hardly distinguishes it from physics or economics. Geography holds in common with the other sciences both its content and its method.[3]

(b) Again, there is little indication in Kant that geography involves a fundamentally different perspective of the study of nature, aside from the fact that he draws a basic distinction between geography and history of nature on the grounds that the former is concerned primarily with spatial matters whereas the latter is concerned primarily with temporal matters. For Kant, however, geography is not a non-temporal science any more than history of nature is a non-spatial science, and moreover, geography has no special claim to space since all other natural sciences are inevitably concerned with spatial matters. The view that geography involves a peculiarly spatial perspective, and can be differentiated from other sciences on such grounds, is to be found principally in the Hettner-Hartshorne classification of sciences to be discussed later.

(c) The idea that we can divide sciences into broad classes, such as law-finding, "typological," and fact-finding, is foreign to Kant's whole approach to science. All sciences, in order to qualify as sciences, must be law-finding, even if levels of generality of laws can be established, since all strictly scientific statements must be exhibited in their conformity to universal a priori principles of human understanding. Thus, for Kant, a strictly fact-finding discipline could not qualify as science, but would remain a part of "common knowledge," i.e., a discipline that did not seek to establish relations among things. Nevertheless, a basic, even if not an absolute, distinction between law-finding and fact-finding sciences is widely recognized today among philosophers of science. But their notion of a fact-finding science differs from Kant's idea of "common knowledge," since "fact-finding" should not convey the idea of the discovery of isolated facts. The mere aggregation of isolated facts cannot constitute a science. A prerequisite, then, of a fact-finding science is that the facts be related and explained. But in the case of a fact-finding science, since the relations

[3]Lukermann, "Geography as a Formal Intellectual Discipline and the Way in which It Contributes to Human Knowledge," p. 167.

222

established are often not deducible from more general statements, they frequently constitute a unique or individual or
non-repeatable configuration of events. History comes to mind
most readily as the example par excellence of a fact-finding
science. Yet considerable controversy surrounds the distinction. For instance, both Nagel[4] and Hartshorne[5] have pointed
out the even the most advanced law-finding sciences contain
predominantly fact-finding segments. On these grounds alone,
therefore, the distinction should not be regarded as an absolute
one for purposes of distinguishing classes of sciences. In
addition, Carl Hempel[6] has argued that a class of "typological"
sciences does not really exist as independent, but rather represents a sort of incomplete type or half-way house between
the other two classes. In short, various sciences, in terms
of the rigour of their explanations, can perhaps best be regarded as forming a sort of continuum, ranging from the most
exact parts of the physical sciences to debatable historical explanations of the actions of an individual historical actor.

Furthermore, difficulties surround the question of what
precisely constitutes a law. For instance, a prominent sociological theorist has denied that there are any sociological
laws on the grounds that sociology has no theories comparable
to those of physics; since laws are statements "derivable" from
theories, it follows that sociology has no laws:

> The second type of sociological generalization, the so-called scientific
> law, differs from the foregoing [the empirical generalization] in as much
> as it is a statement of invariance derivable from a theory. The paucity of
> such laws in the sociological field perhaps reflects the prevailing bifurcation
> of theory and empirical research. Despite the many volumes dealing with
> the history of sociological theory and despite the plethora of empirical in-
> vestigations, sociologists (including the writer) may discuss the logical
> criteria of sociological laws without citing a single instance which fully sat-
> isfies these criteria.[7]

On the other hand, Nagel[8] includes in his classification of basic
types of laws "determinate property" laws, which are hardly
derivable from theory or even require theory.

[4]The Structure of Science, pp. 548-549.

[5]The Nature of Geography, p. 379.

[6]Aspects of Scientific Explanation, chap. vii, "Typological Methods in the Natural
and the Social Sciences," pp. 155-171.

[7]Merton, Social Theory and Social Structure, p. 96.

[8]The Structure of Science, pp. 75-76.

Although the majority of geographers would probably agree that their science is still primarily a fact-finding one, the issue appears to have little precise bearing on the question of the specific place of geography in a classification of the sciences. Whether geography is primarily law-finding, "typological, " or fact-finding does not provide a sufficient criterion for differentiating it from other sciences, since a number of these other sciences also would each fall primarily into one of the three classes.

Besides Kant's classification, five positions concerning a place for geography in a classification of the sciences are worthy of some consideration: (a) law-finding science of spatial relations; (b) "naively given sections of reality"; (c) the Fenneman-Taylor classification; (d) the Hettner-Hartshorne classification; and (e) synoptic.

KANT'S CLASSIFICATION OF THE
SCIENCES, AND GEOGRAPHY AS A LAW-
FINDING SCIENCE OF SPATIAL RELATIONS

Kant's classification of the sciences, and the place of geography in that classification, was outlined in an earlier chapter. Perhaps the most interesting issue raised by that classification, from a contemporary point of view, is Kant's idea of "levels." But the idea of "levels" is a difficult one. As opposed to a thoroughgoing "unity of science, " it at least has the negative advantage that science has as yet been unable to achieve such a unity, in that we cannot, for instance, deduce the statements of the biological and social sciences from those of the physical sciences. Kant apparently finds a distinct "level" for geography. It was argued that there is a sense in which geomorphology as a physical science can be regarded as operating at a different level than physics, and that statements in biogeography and economic geography may also be thought of as statements that occur at a different level than do some other statements of biology and economics, since they are often not, in any strict sense, deducible from the latter. Still, the idea of "level" does not appear to achieve independence for geography since it cannot "constitute its peculiarity. " A difference in "level" is not a sufficient criterion for differentiating geography from other sciences; this is partly due to the growth of science since Kant's day that has obliterated his exclusive category of geography, and partly due to scientific growth that permits statements to appear at a number of different levels within a single science. In biology, for instance, the statements of biogeography and "natural history" can hardly be regarded as at the same "level" as more general statements that have been arrived at in cytology and physiology, and are certainly not deducible from these latter statements. Yet, the statements of biogeography and "natural history" are as much biological as those of cytology and physiology, since the central focus of investigation is still biological. The mere addition of space or time in such studies adds nothing specific by way of non-biological content.

The point of view that geography is a law-finding science of spatial relations has been discussed in considerable detail in the preceding two chapters. In general, it has not initiated explicit contemporary discussion of the place of geography in a classification of the sciences, aside from the statement in The Science of Geography, that science is concerned with "three great parameters . . . space, time, and composition of matter," of which "space in time" is assigned to geography. However, the view that geography involves a peculiarly spatial perspective, and can be differentiated from other sciences on that basis, has been most fully developed in the Hettner-Hartshorne classification of sciences. But the point of view now under consideration, that geography is a law-finding science of spatial relations, appears to relate more closely to Kant's concept of the matter than do the views of any other group of contemporary geographers. In addition, it does have other implications respecting geography's place among the sciences. The aim of geographers of this persuasion is to make of their discipline a law-finding science, and the key concept that occurs repeatedly in pertinent literature is the concept "model." As we have noted, an important procedure in geography for Kant is the employment of "analogies." Used in its restricted sense of an analogue, a "model" implies the mergence of geography with other disciplines, since, if a structural isomorphism can be established between apparently disparate branches of knowledge, then the two branches can be merged in the sense that explanation can proceed on a common basis. Unfortunately, "model" is one of

225

the most confused and bewildering concepts presently employed in geographical, as well as in much social scientific, literature.

One reads, for example, of "descriptive" models, "explanatory" models, and "predictive" models. Basic concepts in philosophy of science such as "percept," "hypothesis," "law," and "theory" are frequently replaced by the single concept "model." In fact, it sometimes seems that much of the vocabulary of the philosophy of science has disappeared into the concept "model." Much of contemporary geography is thus in danger of becoming pseudo-science backed up by pseudo-philosophy of science. The situation has become so bad that one geographer has found it necessary to create a "model of models" to keep track of all the other models.[9] The models contained within the "model of models" appear to range all the way from initial abstractions from reality to general theories. Although Chorley quotes the philosopher Max Black to the effect that "it is important to differentiate between models and theories," we find him a few pages later talking about the Newtonian "model" (pp. 128, 136).

Unfortunately, there is little agreement among philosophers of science themselves as to how the concept "model" should be defined and used. One writer, for instance, has discovered nine different "functions" that models perform in empirical science, and he chides R. B. Braithwaite for using the concept in only one restricted sense.[10] Stephen Toulmin[11] sanctions the use of "models" as precise explanatory devices. He points out that in the science of optics it is common procedure to explain what has happened by means of geometrical models. By means of the model, the source of light and light travelling down a dotted line, etc. can be depicted. The geometrical model employed in optics can stand as an exact representation of the structure of the optical situation being investigated. Toulmin even declares that, for explanatory purposes, "a theory is felt to be entirely satisfactory only if the mathematical calculus is supplemented by an intelligible model" (pp. 34-35).

[9]Richard J. Chorley, "Geography and Analogue Theory," Annals, AAG, LIV, No. 1 (1964), 127-137; for a diagram of Chorley's "model of models," see p. 129.

[10]Leo Apostel, "Towards the Formal Study of Models in the Non-Formal Sciences," in The Concept and the Role of the Model in Mathematics and Natural and Social Sciences, ed. by B. H. Kazemier and D. Vuysje, pp. 1-37, especially pp. 1-3.

[11]The Philosophy of Science: An Introduction, pp. 31-35.

This claim, however, is disputed by Arthur Pap[12] on the grounds that the principal role of scientific theories is not to produce "pictures of reality" but rather to serve as postulates for explanatory-predictive purposes. Models, as "pictures of reality," have no place in strictly scientific explanation. Max Black[13] permits extensive use of the concept, since he recognizes "miniature," "scale," "analogue," "mathematical," and "theoretical" models as legitimate types.

Despite the growing tendency among philosophers of science, as well as among geographers and social scientists, to attach ever wider meanings and uses to the concept "model," it seems preferable to use it in the narrow and specific sense in which it is employed by such philosophers of science as R. B. Braithwaite, Ernest Nagel, and May Brodbeck, and in the specific sense in which Kant appears to use the concept "analogy." Dr. Brodbeck's dictum, "'One thing, one word' is still a good idea,"[14] is especially necessary to avoid the confusion and philosophical naivety that results from using the concept "model" as a substitute for such concepts as "percept," "law," "hypothesis," and "theory." In cases where widely accepted and recognized terminology exists, the indiscriminate use of the concept "model" cannot be excused, and it presupposes an almost total lack of acquaintance with the language of the philosophy of science.

There are two root meanings of the word "model" in ordinary English.[15] In its first meaning, a model is a copy, replica, or miniature. We speak of a model of the new city hall, or of a model aeroplane. In its second meaning, a model is a norm or ideal, as when we speak of a model student. In the first case, the better the model, the more precisely it depicts the original. In the second case, the model is an amalgam or idealization of all the key qualities that are conceived as making up the ideal student. The first root meaning is the basis of scientific models in the sense of analogues; the second is the basis of "ideal types."

[12]An Introduction to the Philosophy of Science, pp. 355-356.

[13]Models and Metaphors, chap. xiii, "Models and Archetypes," pp. 219-243.

[14]May Brodbeck, "Models, Meaning, and Theories," in Symposium on Sociological Theory, ed. by Llewellyn Gross, pp. 373-403; quote from p. 381.

[15]Ibid., pp. 373-374.

Although the terms "model" and "analogy" are often used interchangeably, a distinction between them can perhaps be drawn.[16] An analogy may be defined as a concept or device which is taken from one field of scientific study or area of human experience, and applied in some manner, usually for purposes of elucidation, to another scientific field or area of human experience with which the analogy appears to have some similarity, but with which it is not isomorphic. Analogies, in this strict sense, have only a limited place in scientific procedures, aside from their heuristic value in the initial stages of a scientific investigation to suggest possible alternative hypotheses. However, they also have some value as aids to understanding, or as loose explanatory tools. Thus the famous devices employing clocks, elevators, trains, and ships, used by Einstein and others to help explain the theory of relativity to the lay public, are properly speaking analogies and not models. The theory of relativity is not constructed out of or concerned with, nor is it like, the relations peculiar to clocks, elevators, trains, and ships. These are analogies, employed for the purpose of giving laymen some crude insight into the workings of the theory of relativity. On the other hand, the condition of isomorphism is vital to the concept of a "model." A model may be defined as a device, concept, or theory borrowed from one scientific field and applied in another which possesses a formal structure similar to structures under investigation in that field. The concept presupposes that the field from which the theory is borrowed is more advanced than the field for which it is borrowed, since there is no point in attempting to understand one thing in terms of something else that is less clearly understood. However, as Nagel points out, "It must nevertheless be acknowledged that there is no way of telling in advance whether a given model will prove to be an obstacle to the fruitful development of theory, since it is usually only after a model has been tried that one can tell which of its features suggest inquiries leading into blind alleys and which are heuristically valuable."[17] Thus, there is no way of telling before-

---

[16]This distinction implies in turn that a further distinction can be made between the concepts "analogue" and "analogy." If "analogue" is taken as a synonym for "model," then "analogy" can be reserved to designate something more vague.

[17]The Structure of Science, pp. 115-116.

hand whether an analogy or a model has been employed, since the condition of isomorphism cannot be presupposed but must be demonstrated. Hence it must be admitted that the practical efficacy of the distinction is weakened.

Braithwaite distinguishes between theories and models in the following manner:

A theory and a model for it . . . have the same formal structure, since theory and model are each represented by the same calculus. There is a one-one correlation between the propositions of the theory and those of the model; propositions which are logical consequences of propositions of the theory have correlates in the model of these latter propositions in the theory, and vice versa. But the theory and the model have different epistemological structures.[18]

Although not a firm believer in the use of models, since his ideal for science is that of a purely deductive system, Braithwaite is here setting out a necessary, even if somewhat rigorous, condition for their use in science. He gives several examples of models that have been employed in scientific investigation; for instance, atoms may be thought of as miniature solar systems with electronic planets revolving around a protonic sun (p. 93). Presumably, the solar system and atoms possess the same formal structure. Atoms can therefore be thought of as if they were solar systems because in a formal sense they are like a solar system in some respects. The model has utility as a heuristic device, but it must be clearly distinguished from theory and from the subject matter to which it is applied. Models, however, may have serious shortcomings and disadvantages. As Braithwaite points out: "Hydrogen atoms are not solar systems; it is only useful to think of them as if they were such systems if one remembers all the time that they are not. The price of the employment of models is eternal vigilance" (p. 93). In addition, Nagel[19] points out that often several models are known for the same theory. Furthermore, basing his argument on a famous proof by Henri Poincaré, he demonstrates that an infinite number of models is logically possible. Hence, the mere possession of a model may mean nothing scientifically.

[18]Richard B. Braithwaite, Scientific Explanation, p. 90.
[19]The Structure of Science, pp. 116-117.

The second root meaning of the word "model" in ordinary discourse gives rise to "ideal types." The term "ideal type," however, is frequently replaced, especially in geographical and social scientific literature, by the term "model." This is understandable, since the term "ideal type" is awkward, is not as neat as the term "model." In addition, there is ample precedent in ordinary discourse for the use of the term "model" in its "ideal type" sense. Nevertheless, if the term "model" is to be used in place of the term "ideal type," and this usage is well entrenched at present in much geographical and social scientific literature, then the term "model" in its first sense of an analogue must be clearly distinguished from its second sense of an "ideal type." Essentially, the difference is that "models" as analogues presuppose a condition of isomorphism and thus a borrowing of certain structural features from one area of scientific investigation and their application to another such area, whereas "models" as "ideal types" presuppose no such condition of isomorphism, but rather consist of logically constructed abstractions from reality.

The nature of "ideal types" has perhaps been best set forth by Max Weber.[20] The "ideal type" is constructed from empirical insights, out of the fabric of the subject matter being studied. Weber constructed a variety of "ideal types," ranging from capitalism, to bureaucracy, to the mediaeval craft guild, to the caste system of India, etc.[21] The "ideal type" is constructed by exaggerating and distorting what appear to be the main tenets of the matter under investigation into a logically constructed closed system that ideally represents the reality

[20] The Methodology of the Social Sciences, pp. 89-102; and The Theory of Social and Economic Organization, pp. 81-102.

[21] J. W. N. Watkins calls attention to the fact that Weber held two conceptions of what an "ideal type" should be—holistic and individualistic. Holistic "ideal types," as set forth by Weber in The Methodology of the Social Sciences, consisted of "detecting and abstracting the overall characteristics of a whole situation, and organizing these into a coherent scheme." Weber's later view, as set forth in The Theory of Social and Economic Organization, consisted of "placing hypothetical, rational actors in some simplified situation, and in deducing the consequences of their interaction." In this context, we are concerned with "ideal types" in the first sense, since they are more pertinent to the study of large scale social and economic phenomena. "Ideal types" in the second sense are more pertinent to the study of individual historical situations. (See Watkins, "Ideal Types and Historical Explanation," in Readings in the Philosophy of Science, ed. by Herbert Feigl and May Brodbeck, pp. 723-743.)

being investigated, assuming that it operated perfectly. The investigator, presumably, then has a clear, unobstructed view of the key features present. The "ideal type" can be employed as a standard against which to compare the distortions present in reality, in various concrete examples of the "ideal type." One can then proceed to study how reality departs from the "ideal type," and attempt to discover the causes for marked deviations, or the reasons for differences among concrete examples that can be subsumed under the "ideal type."

There are, of course, many dangers inherent in the use of "ideal types." It is very difficult to tell, for instance, whether one has succeeded in isolating the main components from the welter of material which one may be studying. There would appear to be no objective standards against which to check such a procedure. In addition, it is difficult to determine precisely what the relations among the abstracted major components are or ought to be in the "ideal type." The "ideal type" has the added danger of focussing attention on material that will tend to support it. As a result, an "ideal type" can become out of date, in the sense that it ceases to be a reasonable reflection of the reality in whose investigation it is supposed to assist, without the investigator being aware of the fact. Or, one may become so obsessed with the internal logic of "ideal type" construction, that the "ideal types" become more and more remote from reality. There is, of course, nothing new about these drawbacks to the employment of "ideal types." For instance, it is now more than a century since von Thünen first pointed out the weaknesses of his own employment of "ideal types"— weaknesses, moreover, that he was amazed his critics had not raised and discussed. As he remarks in The Isolated State:

The abstraction from reality without which we cannot come to any scientific knowledge has several dangers, namely: (1) We separate in thought what is in fact mutually interrelated. (2) Our conclusions rest upon assumptions of which we are not clearly conscious and which we therefore do not make expressly, and we then consider as generally true what is true only under these specific assumptions. The history of economics gives us many striking examples.

Among the assumptions mentioned expressly in the first volume or quietly assumed, there are two which require special examination and clarification: (1) The soil in the plain of the Isolated State is not only originally of equal fertility but after cultivation, with the exception of the first circle, equal fertility remains in regard to the ability of soil to grow plants in all parts of

231

the Isolated State howsoever different the prices of grain may be. (2) The diligence exercised in agriculture, in plowing, harvesting, or in anything else, is everywhere the same, whether the bushel of rye is worth one half or one and one-half talers. Now, we must put the rationality of economic activity in the first place, and subordinate everything else.

The question arises of itself: "Are both these assumptions consistent with rational management?" To that I must answer no. The reasons for this answer must be further developed. From this point of view, Volume 1, which does not justify this, could have been attacked and would have been attacked if the book had received criticism in the spirit of the work itself.[22]

"Ideal types," because of a lack of adequate theory, have been most frequently employed in social scientific research; in fact, the social sciences can be regarded as the domain par excellence of "ideal types." Although economics is generally regarded as the most advanced of the social sciences, it has probably made more extensive use of models of the "ideal type" variety than has any other social science. For instance, economic "models" concerning perfect competition, monopoly, demand and supply, perfect equilibrium, etc., are essentially "ideal types." Perfect competition does not exist in the real economic order, nor do supply and demand operate with the perfect rationality of theory. Rather, these are rational constructs, created out of the fabric of a real economy, where they can be observed in operation in imperfect form. Moreover, many of the so-called "theoretical" constructs of contemporary geography are essentially "ideal types." For instance, William Bunge regards "central place theory" as a major theoretical breakthrough for geography. Yet, Walter Christaller, the acknowledged leader of geographical concern with "central place theory," has stated explicitly that it consists essentially of "ideal type" constructs in Max Weber's sense of the term:

We also made extensive use of the geographical method when we laid out the theoretical pattern of central places over the surface of the globe, and then compared the actual circumstances with these ideal circumstances (the "ideal types" of Max Weber).[23]

[22]Johann Heinrich von Thünen, The Isolated State, trans. by Bernard W. Dempsey in his The Frontier Wage, pp. 197-198.

[23]Christaller, Central Places, p. 200; parentheses are in the text. Although Christaller evidently regards the theoretical basis of his methodology as derivable from Max Weber (p. 9), he nevertheless appears to regard "ideal types" as "theories" that explain reality at least in part, and moreover, he states that geographical and historical factors that depart from "theory" or are not explained by it "cannot be cited

If "ideal types" are not theories, one might then inquire as to the differences between the two. To begin, no hard and fast distinction can be drawn between the physical sciences and the social sciences with respect to the employment of "ideal types," since the less theoretically developed branches of physical science certainly employ them. For instance, in the absence of "tight" theory, the idea of a "standard atmosphere" as employed in meteorology is exceedingly useful as an organizational and exploratory device that facilitates comparative studies. Carl Hempel argues that there are two respects in which theories differ from "ideal types." Using economics, the most advanced of the social sciences, as his example, he states:

In two important respects, however, idealizations in economics seem to me to differ from those of the natural sciences: first of all, they are intuitive rather than theoretical idealizations in the sense that the corresponding postulates are not deduced, as special cases, from a broader theory which covers also the nonrational and noneconomic factors affecting human conduct. No suitable more general theory is available at present, and thus there is no theoretical basis for an appraisal of the idealizations involved in applying the economic constructs to concrete situations. This takes us to the second point of difference: the class of concrete behavioral phenomena for which the idealized principles of economic theory are meant to constitute at least approximately correct generalizations is not always clearly specified. This of course hampers the significant explanatory use of those principles: an ideal theoretical system, as indeed any theoretical system at all, can assume the status of an explanatory and predictive apparatus only if its area of application has been specified; in other words, if its constituent concepts have been given an empirical interpretation which, directly or at least mediately, links them to observable phenomena.[24]

directly as proof against the validity of the theory" (pp. 4-5). However, Christaller remarks that the general statements of social scientific "theories" should be designated as "tendencies" rather than as "laws," since "they are not so inexorable as natural laws" (p. 3). There is no evidence, however, to suggest that Weber himself regarded "ideal types" as theories, as embodying law-like statements, or as explanatory in any strict sense of that term. On one occasion he remarked that "it [the "ideal type"] is no 'hypothesis' but it offers guidance to the construction of hypotheses. It is not a description of reality but it aims to give unambiguous means of.expression to such a description" (The Methodology of the Social Sciences, p. 90). In other words "ideal types" are essentially heuristic devices that at best may serve as guide-lines to the construction of hypotheses, and as aids in abstracting from reality those major components that may eventually lead to an adequate scientific "description" of reality. Moreover, assistance in analyzing and understanding "unique" or "individual" configurations of events appears to have been the special purpose that Weber had in mind for his "ideal types" (pp. 93, 100-101).

[24]Hempel, Aspects of Scientific Explanation, pp. 169-170.

In brief, "ideal type" constructs lack strict scientific justification in that they are intuitive, i.e., not deducible from more theoretical statements possessing empirical warranty, and in that they are often not clearly enough separated from closely related matters that affect them, or are not specified in precisely empirical fashion. In addition, as Hempel indicates, "ideal type" constructs are not strictly speaking explanatory, in that no particular statements can be deduced from theoretical statements possessing empirical warranty. In other words, since "ideal type" constructs are idealizations of empirical reality, then, in most instances, reality will depart from the "ideal type" to varying degrees. One is always entitled to ask why, and also why one concrete example loosely subsumable under the "ideal type" may differ from another. However, even if one were to discover a concrete case that exactly fitted the "ideal type" one would still be entitled to ask why it did. On the other hand, there is no reason for asking the question "Why?" with respect to a particular case that can be subsumed under, or deduced from, a theoretical statement in physics that possesses empirical warranty, unless one is prepared to embark on an exploration of metaphysical or theological considerations. Thus, with respect to the explanatory status of "ideal types," we appear to be driven to the position that we must either press on to the establishment of theoretically rigorous explanations or fall back upon historical explanations, either intuitive or statistical.[25] However, since "ideal types" represent explanatory halfway houses, or paths to explanation, or only partially developed explanatory systems, the possibility exists that their further refinement may eventually lead to the establishment of theory.[26] Thus, even aside from their present utility, the continued employment of "ideal types" in social scientific and geographical research is justified and should be encouraged.

[25]For enlargement of this point, see Watkins, "Ideal Types." Watkins himself opts for intuitive historical explanation.

[26]This is essentially Hempel's view of the scientific status of "ideal types" (Aspects of Scientific Explanation).

"NAIVELY GIVEN SECTIONS OF REALITY"

Carl Sauer has argued that sciences are differentiated from
one another principally on the basis of representing "naively
given section[s] of reality."[27] Thus, "area or landscape is
the field of geography, because it is a naively given, important
section of reality, not a sophisticated thesis." More recently,
Lukermann has also put forth the same claim, and likewise
chooses "area" as geography's "naively existing . . . thing it
is going to investigate."[28] Sauer and Lukermann come closest
to attempting to find a place for geography among the sciences
on the basis that it studies a distinctive "object." By "naively
given," Sauer appears to mean something that is immediately
or directly given to us in ordinary experience. But this criterion can hardly serve as the basis for a complete classification
of the sciences, despite the fact that some sciences can be
thought of as studying "naively given sections of reality." For
instance, plants, animals, and rocks are things that we can
immediately identify and differentiate in ordinary sense experience, and thus botany, zoology, and geology can be regarded
as sciences that rest ultimately on a "naively given" basis.
Similarly, on a clear night, one can look up at the heavens and
see a multitude of stars, and thus the science of astronomy can
also be thought of as studying a "naively given," albeit very
large, section of reality. On the other hand, since they are
highly abstractive and experimental, there is no sense in which
universal sciences such as physics and chemistry can be regarded as "naively given." History and geography present
special difficulties in this regard. Although we can, over time,
be aware of changes in our own lives, and of changes in the society around us, I doubt that this type of experience leads to
any immediate awareness of the past that preceded our own
lives, and hence it is very difficult to think of history as studying a "naively given section of reality." And although we can
be immediately aware of spatial relations among things in the
world around us, I seriously doubt that we have any immediate
awareness of the existence of "areas" or "regions" of the

[27]"The Morphology of Landscape," in Land and Life, ed. by Leighly, p. 316.

[28]"Geography as a Formal Intellectual Discipline and the Way in which It Contributes
to Human Knowledge," 167.

earth's surface. Hence, geography can hardly be regarded as studying a "naively given section of reality" in the sense in which botany can perhaps be so regarded.

THE FENNEMAN-TAYLOR CLASSIFICATION

This classification can best be introduced by means of a simplified illustration (see Fig. 2).[29] The portion of the diagram contained within the main circle represents the domain of geography, whereas the portion of the diagram outside the main circle represents the domains of other sciences. Reality beyond the main circle is studied by a variety of specialized disciplines, only eight of which are depicted. These specialties are represented by the series of smaller circles which dissect and overlap the main circle. Thus, since these smaller circles overlap the larger one, it follows that elements of geography are contained within each of the specialized disciplines that are represented by the smaller circles. In other words, although these portions of the larger circle contained within the smaller circles are geographical in the broad sense of that term, in that they represent a concern with environmental factors, or with spatial distribution and spatial interaction over the surface of the earth, they are logically subsumable under the various specialized non-geographical sciences. This interpretation of the diagram is clearly indicated by Fenneman when he remarks: "This diagram expresses the fundamental conception that sciences overlap and that each one of the specialized phases of geography belongs equally to some other science."[30] Fenneman puts the issue even more bluntly when he states that "it may be well to add in plain English that the

---

[29]The diagram has been adapted from Nevin M. Fenneman, "The Circumference of Geography," reprinted in Outside Readings in Geography, ed. by Fred E. Dohrs, Lawrence M. Sommers, and Donald R. Petterson, p. 3; and from Griffith Taylor, "Introduction: The Scope of the Volume," Geography in the Twentieth Century, p. 18. The diagram is an adaptation, since it eliminates such illogicalities as setting historical geography under political science. In addition, the central portion of the diagram, which Fenneman labels "geography" or "regional geography," is labelled "human environmental" by Taylor. Nevertheless, the text makes clear that what Taylor calls "human environmental" is "world regional geography" in Fenneman's (and in the customarily recognized) sense of the concept.

[30]"The Circumference of Geography," p. 3.

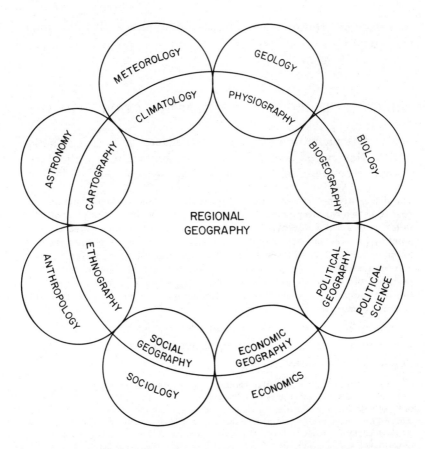

Fig. 2. The Fenneman-Taylor Classification of Sciences

one thing that is first, last, and always geography and nothing
else, is the study of areas in their compositeness or complexity, that is regional geography" (p. 6). Taylor is less clear
respecting the logical status of these specialized branches of
geography, but he appears to endorse Fenneman's position
when he remarks that, essentially, "Geography correlates the
specified branches of the four 'Environmental' Sciences with
the four 'Human' Sciences."[31] Hence, as correlative, and as
an independent discipline, geography is essentially regional
geography. However, since regional geography correlates

[31]Griffith Taylor, "Introduction: The Scope of the Volume," p. 18.

data derived from more specialized disciplines, it is a synoptic science.

THE HETTNER-HARTSHORNE CLASSIFICATION

Again, this classification can be introduced by means of a simplified illustration (see Fig. 3).[32] An explanation of the diagram can be given in Hartshorne's own words:

> The planes are not to be considered literally as plane surfaces, but as representing two opposing points of view in studying reality. The view of reality in terms of areal differentiation of the earth surface is intersected at every point by the view in which reality is considered in terms of phenomena classified by kind. The different systematic sciences that study different phenomena found within the earth surface are intersected by the corresponding branches of systematic geography. The integration of all branches of systematic geography, focussed on a particular place in the earth surface, is regional geography.[33]

Furthermore, a third plane intersecting the other two, and representing the temporal or historical perspective, could be included in the diagram.[34]

[32] The diagram is taken from Hartshorne, The Nature of Geography, p. 147. Hartshorne, in turn, apparently took the diagram from Alfred Hettner, since he refers to it as "this figure of Hettner's, which could be derived also from Kant" (p. 146). Although I have been unable to discover the diagram in Hettner's writings, aspects of those writings certainly sanction such a diagram: "Reality is simultaneously a three-dimensional space, which we must examine from three different points of view in order to comprehend the whole; examination from but one of these points of view alone is one-sided and does not exhaust the whole. From one point of view we see the relations of similar things, from the second the development in time, from the third the arrangement and division in space. Reality as a whole cannot be encompassed entirely in the systematic sciences, sciences defined by the objects they study, as many students still think. Other writers have effectively based the justification for historical sciences on the necessity of a special conception of development in time. But this leaves science still two-dimensional, we do not perceive it completely unless we also consider it from the third point of view, the division and arrangement in space" (Hettner, Die Geographie, pp. 114-115; trans. by Hartshorne, The Nature of Geography, p. 140). It should also be pointed out that Hettner (Die Geographie, p. 114), basing his position on Comte, makes an initial distinction between "abstract" and "concrete" sciences. In his classification, he is concerned only with relations among the "concrete" sciences and hence only with the relations between geography and other "concrete" sciences.

[33] The Nature of Geography, p. 147.

[34] This is certainly sanctioned by Hartshorne when he remarks: "The comparison with history, then, emphasizes for us the special character of geography as an integrating science cutting a cross-section through the systematic sciences" (ibid., p. 146).

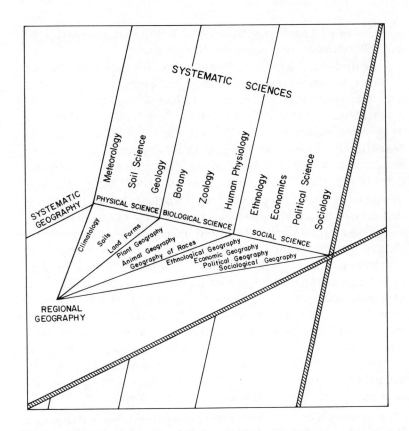

Fig. 3.   The Hettner-Hartshorne Classification of Sciences

This issue of the logical independence of a class of spatial
or geographical sciences has been debated at considerable
length in previous chapters. In brief, the outcome of the debate
was that there are no logical grounds upon which to establish
an independent class of spatial or geographical sciences, any
more than there are grounds upon which to establish a class of
temporal or historical sciences. Rather, these "classes" of
sciences are logically a part of the content of the various "subject matter" sciences. Thus, the Hettner-Hartshorne classification appears to end up in the same position as the
Fenneman-Taylor classification, namely, that the core of ge-

ography is regional geography, and hence that geography is only classifiable as a synoptic discipline.

Of historical interest, however, is the fact that this view of the relations between geography and other sciences has often been attributed to Kant, or at least has been regarded as originating with, or traceable to, Kant. For instance, Hettner, [35] as we have seen, remarks that his views on the topic had been "excellently" expressed by Kant in his lectures on physical geography; and Hartshorne, as we have also seen, believes that his diagram "could be derived also from Kant."[36] Ample evidence has been produced in this work, however, to suggest that this is not, and cannot be, Kant's view of the matter.

Especially unfortunate is the idea, expressed by both Hettner and Hartshorne, that geography can be divided into "systematic" and "regional" parts. This distinction implies that "regional" geography must be non-systematic. On Kantian grounds, however, if it is non-systematic it is non-scientific, since all sciences are necessarily "systems." For Kant, regional geography is merely another way of doing geography systematically, and moreover, all geography, by its very nature, is "regional." Hartshorne, however, has recently modified his position somewhat. The following statement may be taken as a summary of his present view of the issue:

In sum, geography cannot be considered as divided between studies which analyze individual elements over the world and those which analyze complete complexes of elements by areas. The former are logically a part of the appropriate systematic sciences, the latter simply cannot be carried out. All studies in geography analyze the areal variation and connections of phenomena in integration. There is no dichotomy or dualism, but rather a gradational range along a continuum from those which analyze the most elementary complexes in areal variation over the world to those which analyze the most complex integrations in areal variation within small areas. The former we may appropriately call "topical" studies, the latter "regional" studies, provided we remember that every truly geographic study involves the use of both the topical and the regional approach.[37]

Since "studies which analyze individual elements over the world . . . are logically a part of the appropriate systematic

[35]Die Geographie, p. 115.

[36]The Nature of Geography, p. 146.

[37]Perspective on the Nature of Geography, pp. 121-122. Hartshorne himself points out in a footnote, p. 122n, that his present "viewpoint . . . differs markedly from that presented in The Nature of Geography."

sciences," it would appear that the so-called "systematic" aspects of geography, which study the spatial distribution and spatial "integration" of these "individual elements," and which Hartshorne formerly regarded as logically a part of geography, are no longer to be so regarded. If this is the case, then "regional" geography in its former sense becomes the whole of geography. However, the distinction which Hartshorne now draws between "topical" and "regional" then becomes quite unclear. For, if all geographical studies form a "continuum" ranging from the "topical" to the "regional," it becomes impossible to understand why "every truly geographic study involves the use of both the topical and the regional approach." The use of the term "topical" in this context seems to involve a throwback to the idea that there are "systematic" aspects of geography which are concerned with an analysis and understanding of "individual elements," and which the geographer must thoroughly comprehend before he can engage in regional studies. Nevertheless, despite these apparent modifications of his view, Hartshorne still holds to his original position respecting the place of geography in a classification of the sciences, and still regards that position as derivable, basically, from Kant and von Humboldt.[38]

The origin of the Hettner-Hartshorne view of geography's place among the sciences appears to stem from the influence on Hettner of neo-Kantian philosophy, and from his own modifications of that philosophy. As quoted earlier, Hettner remarks that: "Other writers have effectively based the justification for historical sciences on the necessity of a special conception of development in time."[39] This passage appears to involve a reference to the neo-Kantian distinction, popular in Hettner's day, between cultural-historical sciences and natural or physical sciences. Extending the notion quite logically to encompass space, Hettner then proceeds to recognize a distinct class of geographical or spatial sciences. Nevertheless, he refuses to draw an absolute distinction between nomothetic and idiographic sciences,[40] thus following

---

[38]Perspective on the Nature of Geography, p. 180.

[39]Die Geographie, p. 114.

[40]Hettner himself points out that, expressly in the case of Rickert, no such hard and fast distinction can be drawn (ibid., pp. 112-113).

his mentors, Rickert[41] and Windelband, [42] to whom he refers as "two outstanding philosophers."[43] And so, for Hettner, the possibility exists that geography can be conceived as an independent yet generalizing or law-finding science of spatial relations. However, as we have seen, this position is logically untenable.

Since the Hettner-Hartshorne classification advocates the independence of historical and geographical classes of science, this seems to be the best place to raise the issue of explanation. If it can be established that there are distinctive forms of historical and geographical explanation, the case for the independence of history and geography as distinct disciplines would thus be strengthened. Philosophers of science as well as philosophers of history, in increasing numbers, are coming to recognize historical explanation as a distinctive form of explanation, even though different interpretations of the nature of historical explanation continue to be given.[44] If there is a distinctive form of historical explanation, then there are grounds upon which to regard history and historical studies generally as somehow distinct from other types of scientific investigation, even if they do not constitute an entirely distinct class of sciences. The question then arises: Is there a distinct form of geographical explanation?

Advocates of "individualistic" historical explanation adopt the point of view that explanation in human history is fundamentally different from explanation in the law-finding physical sciences.[45] In the advanced physical sciences, an explanation is achieved by subsuming a statement about a particular situation under a more general statement, by exhibiting the statement as an instance of a general law. Historical explanation is achieved by understanding a unique, or individual, or non-

[41]Science and History, pp. xii, 3.

[42]Wilhelm Windelband, Theories in Logic, especially pp. 57-58.

[43]Hettner, Die Geographie, p. 112.

[44]There are, of course, those who take exception to the generalization that a distinctive form of historical explanation exists. For example, Hempel (Aspects of Scientific Explanation, pp. 447-453, and "Reasons and Covering Laws in Historical Explanation," in Philosophy and History: A Symposium, ed. by Hook, pp. 143-163) still regards "historical" or "genetic" explanation as essentially nomological in character.

[45]For a detailed analysis and exposition of this point of view, see William Dray, Laws and Explanation in History.

repeatable action, situation, or configuration of events. The historian achieves this understanding by, in a sense, reliving the past and rethinking the thoughts of some historical actor, in short, by putting himself in the place of the historical actor. Explanation in terms of general laws is explanation from a spectator's point of view; historical explanation is explanation from an actor's point of view. The historian is not concerned with trying to exhibit the unique historical event or action as an instance of some general law; he is concerned with understanding the event or action in and for itself. This is not to deny that at some point in his considerations the historian may not employ generalizations concerning "human nature" as aids to understanding. But the historian is not satisfied with—although, because of lack of evidence, he may at times have to settle for—the form of explanation which maintains that when in situations of type A, historical actors of type B usually or even invariably do actions of type C. Rather, he is interested in understanding, in individual terms, the fact that in particular situation X, historical actor Y did Z.

Philosophers of history or philosophers of science who adopt an external or "holistic" approach to historical explanation tend to deny the uniqueness of historical events, yet may concede that historical explanation possesses some distinctive features. For instance, Nagel[46] contends that explanations of individual historical actions are essentially probabilistic in form, yet is prepared to grant that probability in this context can be only loosely specified. In the case of an explanation of a series of events, following one another in temporal sequence, Nagel argues that the historian often employs probabilistic and even deductive modes of explanation of individual sequences within the series, yet implies that the series as a whole cannot be explained either deductively or probabilistically since there is no apparent logical connection between event one and the final outcome of the series; in other words, too many "brute facts" intervene as required steps in the sequence of events. Nevertheless, Nagel concludes that any particular historical event of this type is analyzable into a sequence of probabilistic explanations. However, whether or not the historical event is analyzable in such a logical fashion, i.e., can be exhibited as

---

[46]The Structure of Science, pp. 551-575.

an instance of such a form of explanation, without doing some injustice to historical explanation, is perhaps still debatable. An important part of the historian's task, in order to achieve a "full" historical understanding of the event, in Popper's opinion, is "to include aspects which [he] does not attempt to explain causally, such as the 'accidental' concurrence of causally unrelated events." Moreover, if "the 'accidental' manner in which these [causal] threads are interwoven"[47] is a necessary aspect of historical understanding, it is difficult to see how the understanding of an historical event of this type is strictly reducible to a sequence of probabilistic explanations, without leaving aside an important aspect of the historian's account.

Historical explanations are not peculiar to history or to the social sciences. For instance, T. A. Goudge[48] argues that historical modes of explanation are essential components of the theory of evolution in biology. And historical explanations, moreover, are certainly employed in geology and geomorphology. In biological and physical contexts, however, the term "historical explanation" cannot mean what it does when employed by those who regard study within human or social contexts as being concerned with actors and with the "inner" side of historical occurrences, since there are no actors, and there is no "inner" side, in the former contexts. In those contexts, "historical explanation" refers to the reconstruction of some past sequence of occurrences which cannot in any strict sense be explained through subsumption under general laws, and which cannot be expressed in any strict sense in the form of law-like statements.

Although it is apparent that considerable controversy surrounds the logical status of historical explanation, there are nevertheless some grounds for regarding that form of explanation as being at least in part distinctive. On the other hand, it would appear that there is no distinctive geographical form of explanation, any more than there are distinctive forms of chemical, economic, or sociological explanation. Geography, in common with a number of other sciences, makes use of a variety of forms of explanation.[49] Geomorphologists in their re-

---

[47] Karl R. Popper, The Poverty of Historicism, p. 147.

[48] The Ascent of Life, pp. 62, 65-79.

[49] These "forms" of explanation—"deductive," "probabilistic," "functional," and "genetic"—are those set forth by Nagel, The Structure of Science, pp. 20-26.

search evidently employ the deductive form of explanation, in that they account for some aspects of the transformation of landforms by subsuming those aspects under general physical laws. Geomorphologists sometimes speak of their discipline as a "pseudo-science," apparently on the grounds that it cannot yet explain in the same manner as can physics:

Even in 1963 it [geomorphology] is still a 'pseudo-science' or a 'quasi-scientific art' and its historian is faced with the portrayal of its early rise both as an artistic perception and as a gradual transition towards a goal of truly scientific explanation which it has so far not attained. [50]

Because of the ever increasing use of a variety of statistical techniques and procedures in geographical research, the probabilistic or statistical form of explanation is often employed. Since geographers are often interested in spatial interaction, it might be argued that the functional form of explanation, i.e., demonstrating how various factors mutually condition the state of some system, is most characteristic of geography. This form of explanation, however, is common in various physical and social sciences. And finally, historical explanation is employed in geography, not only in geomorphology, but also in the study of "places" or "regions" when an attempt is made to account for the present spatial configuration of things on the basis of how that configuration has come to be as it now is, and in historical reconstructions of some past pattern of land use or spatial integration of phenomena.

SYNOPTIC

If geography has no objects of study that are peculiarly or essentially its own, then it would appear that it cannot be classified in the same manner as are a number of other sciences. The outcome of this chapter and of preceding chapters is that geography is, in several respects, a synoptic discipline. The questions then arise: (1) Is a separate class of synoptic sciences possible?; and (2) If so, can geography be distinguished from other synoptic sciences?

[50]Richard J. Chorley, Antony J. Dunn, and Robert P. Beckinsale, The History of the Study of Landforms or the Development of Geomorphology, Vol. I: Geomorphology before Davis, p. xi.

Synoptic sciences, however, present special difficulties for the problem of the classification of the sciences. A. C. Benjamin[51] points out that synoptic sciences cannot be classified on the customary basis that distinguishes sciences according to differences in the objects they study, since this would involve a cross-classification. Nor can the difficulty be resolved by making an initial distinction between special sciences and synoptic sciences, since all sciences, except the most highly specialized, are to some extent synoptic relative to their various parts. And so it would appear that synoptic sciences are of a peculiar kind, not classifiable in the customary manner.

Despite the cross-classificational difficulties involved in establishing a distinct class of synoptic sciences, we nevertheless recognize some kinds of study as being essentially synoptic. These studies are essentially and not just internally synoptic, since they involve the integration of data that are studied by distinct sciences, and not the integration of data that are studied by branches of a single discipline. Thus, plant ecology integrates or synthesizes data derived from botany, geology, pedology, and climatology. But in spite of prima facie evidence for the recognition of a class of synoptic sciences, major logical difficulties arise in the attempt to distinguish clearly one synoptic science from others. On the grounds that any discipline which synthesizes data that are studied by any two sciences is a synoptic science, then a wide range of synoptic disciplines are logically distinguishable from one another. In fact, the combinations involved result in the recognition of far more synoptic sciences than there are essentially subject-matter sciences. But what of the status of synoptic disciplines that synthesize data derived from more than two subject-matter sciences? Here, the permutations that result can become very large indeed, especially since synthesis of data can go on among sciences at a number of different levels. Do we then not end up with a whole hierarchy of synoptic sciences, which culminates in the creation of one super-synoptic science that somehow integrates or synthesizes the data studied by all other synoptic and subject-matter sciences? We appear to be driven to the position that synoptic sciences, no less than

[51] An Introduction to the Philosophy of Science, p. 406.

object-centred sciences, require a recognizable set of objects in order to claim independence as distinct sciences.

Even if we set aside the logical problem of distinguishing synoptic sciences from one another, major pragmatic difficulties are nevertheless involved in distinguishing geography from other synoptic sciences, since it is not the only synoptic science. For instance, varied bioecological studies are well entrenched as elements of biology, and can hardly be claimed as aspects of geography. Although a strong case can be made for regarding geography as the synoptic science that studies man in relation to his physical environment, geographers by no means have the field exclusively to themselves. Medical science, since the time of Hippocrates, has been vitally concerned with studying the relations between human health and various facets of man's physical environment. And moreover, the increasingly important research area that studies world population in relation to world food supply, or more generally, the carrying capacity of the earth, can claim few geographers among its leading exponents. The acknowledged leaders in this field, such as Julian Huxley, John Boyd Orr, E. John Russell, Fairfield Osborne, and Colin Clark, are biologists, economists, agriculturalists, and demographers, not geographers. In fact, perhaps the only geographer in the field who can be considered well-known is L. Dudley Stamp.

When one considers the area of research labelled "human ecology," i.e., the study of the spatial integration of diverse human phenomena, geographical research is virtually indistinguishable from research carried on by a number of social sciences. And if we should eventually succeed in formulating an integrated social science, "human ecology" studies would then become internally synoptic, and would cease to be externally synoptic. In this case, geography as "human ecology" would become little more than the peculiar label we reserved for studies that attempted to integrate spatial aspects of a general social science.

Finally, studies of places and regions are certainly synoptic, and geography can claim some independence because of its preoccupation with such studies. However, whether one would want to accord the label "scientific" to such studies, at least in their present state, is highly questionable.

247

Although it is often argued that synoptic studies are usually less exact than special or topical studies, since they involve the integration of data that are studied by more specialized sciences, this need not always be the case. For instance, the work of C. W. Thornthwaite and his associates on the "water balance," which is basically climatological, and which is synoptic in that it involves a thorough knowledge of some aspects of hydrology, pedology, geology, botany, climatology, and meteorology, has reached a high degree of scientific sophistication. Thus, the possibility exists that other types of synoptic study can be rendered more exact, and hence their scientific justification is assured, even if it appears to be more pragmatic than logical.

X
Conclusion

The answer to the question whether Kant's concept of geography provides an adequate foundation for contemporary geography is, in a word, no. Aside from insufficient or uncertain discussion of some issues, it suffers two major drawbacks. (1) Because of the profuse growth of science since Kant's day, his concept of the limits and scope of geography is inevitably much broader than any contemporary concept can reasonably be; and (2) the distinction between theoretical and empirical sciences, as Kant draws it, is hardly acceptable today, and hence cannot guarantee the logical independence of geography from other disciplines. In addition, some would undoubtedly object to his distinction between outer and inner sense, as a basis for separating geography from other human sciences, on the grounds that, although social science may be conceived as having a peculiarly "inner" perspective, the "outer" perspective cannot logically be excluded from social scientific investigation.

Yet Kant's concept of geography remains extremely suggestive, since all contemporary concepts of the nature of geography are contained, at least inchoately, in his concept. Geography will undoubtedly continue to play a role as a "popular" discipline, much in the manner it did for Kant. Likely to persist is man's perennial interest in "places" and "regions," and geography should continue to fulfill in part this interest through studies ranging all the way from travelogues and travel guides to sophisticated regional monographs comparable to those of Vidal de la Blache and his followers. In addition, it can continue to play a role as a synoptic discipline in the academic sense, through providing an overview of the surface of the earth, the environment of man, which no specialized science can provide. In these roles, however, geography is likely to remain "popular" and to occupy a position peripheral to the main stream of scientific development.

As a science, in the stricter sense of that term, geography's future role appears to be uncertain, especially as a discipline with some claim to independent status. Geographers have attempted to establish their discipline as a science on four different bases: (1) as "environmentalism"; (2) as "ecology"; (3) as a synoptic science in the research sense; and (4) as a science of spatial relations.

(1) Although a case can be made for the claim that geography is the science that studies man in relation to his physical environment, its justification appears to be pragmatic and not logical. Many physical earth scientists and social scientists have eschewed the study of the topic, and have explicitly regarded it as coming within the competence of the geographer. Yet biology, through its interest in the relations of plants and animals to their physical environments, has established a precedent for comparable studies by physical earth scientists and social scientists. Historically, in the main, environmentalism as the concern of geography is a late arrival. Although Strabo definitely regarded such study as geographical, much of the history of Western thought has been dominated by Aristotle, who regarded environmental studies as a part of "politics," i.e., social science generally. Kant was probably one of the first in "modern" times to have assigned environmental studies to geography. In addition, geography has no exclusive claim to the study of man in relation to his physical environment; for instance, throughout the entire history of Western thought, medical science has been vitally concerned with the topic. Finally, environmentalism as the concern of geography, and as an "exact" science, raises the issue of the extent to which a thorough knowledge of some aspects of a wide range of physical and social sciences must be presupposed of the geographer.

(2) "Ecology" does not provide a distinctive concept of the nature of geography. In one sense it is a synonym for "environmentalism"; in another sense, a synonym for "spatial interaction." However, studies which attempt to integrate data derived from a number of more specialized disciplines are also often labelled "ecological." In this area, studies undertaken by geographers are virtually indistinguishable from studies by biologists and social scientists. And furthermore, if one claims that geography is the science of "ecology" in this integrative sense, then one is immediately involved in the problems relating to geography as a kind of super-science.

(3) Again, the notion of geography as a synoptic science in the research sense does not provide a distinctive concept of the nature of geography, since "environmental" and "ecological" studies are clearly synoptic. Nevertheless, one area of geographical research—climatology—appears to fall peculiarly under this rubric. Kraft points out that climatology represents a "common meeting-ground" for geology, the science of the physical surface of the earth, and meteorology, the science of the earth's atmosphere, and falls between the primary interests of each, and thus exhibits this peculiarly synoptic quality. In addition, some areas of research in climatology have attained a considerable degree of scientific sophistication.

(4) Geography, conceived as the generalizing science of spatial relations, has no logical claim to independence, since all sciences are inevitably concerned with spatial relations. Nevertheless, the idea that geography is concerned in some sense with spatial relations has been one of the most persistent concepts of the nature of the discipline throughout the history of Western thought. Hecataeus of Miletus, at the very birth of geography in the sixth century B.C., was possibly the first to have drawn the distinction between a "choros" and a "chronos." Implicit in this distinction was the idea that geography is concerned with the spatial association among things, whereas history is concerned with development and change through time. Kant was the first to have set forth, in sustained discussion, the essential differences between geography and history, although for him no absolute separation of the two is possible. In addition, for Kant, the concept that geography is essentially a science of spatial relations is unquestionably basic. Although geography, as a science of spatial relations, can lay no claim

to logical independence, the fact remains that spatial studies are widely recognized among the biological and social sciences as distinctive areas of research within their respective domains. For instance, the terms "biogeography" and "economic geography" are widely employed to designate relatively distinctive areas of research. This recognition stems from the fact that many biologists and social scientists are not necessarily concerned with the study of spatial relations, in the sense of either spatial distribution of or spatial interaction among phenomena; on the other hand, some biologists and social scientists are concerned with such research, and legitimately so. Yet, as we have seen, biogeographical and economic-geographical statements, for instance, are often not deducible from other statements in biology and economics. This suggests that further empirical research is necessary to obtain generalizations in these areas, and thus supports the position that they have some claim to quasi-independence. However, such research can best be regarded as adjuncts or parts of biological and social scientific research generally, and not as entirely separate areas of research, since such research is necessary for validation of a number of biological and social scientific generalizations.

Since spatial relations are an important area of research and should be fully cultivated, and since this area is often treated only incidentally or peripherally by the biological and social sciences, then support is again lent to the idea that geography as a science of spatial relations has some claim to quasi-independent status. In the case of biology, however, since the question of geographical distribution of biological species or of geographical speciation appears to be such a vital aspect of biological science, and presupposes years of study required to obtain a thorough knowledge of biological species and subspecies, then there is little in the way of even pragmatic justification for regarding such studies as being the concern of the geographer per se. In the case of the social sciences, however, because of their relatively undeveloped state, the matter is less clear-cut, and a stronger case can undoubtedly be made, in relation to them, for geography as a quasi-independent science of spatial relations. Nevertheless, in either case, biological or social, as in the case of geography conceived as "environmentalism," the situation presupposes that the geographer possess a greater knowledge than he ordinarily at present does of the content of biological and social science. This implies an increasing mergence of geography with other disciplines.[1]

Although the arguments set forth in this study lead to the conclusion that, aside from its traditional "popular" commitments, geography as a science will increasingly merge with other disciplines closely related to it, this conclusion need not alarm geographers unduly. At least a quasi-independence seems guaranteed to geography as a science, provided that geographers are prepared to expend the effort to familiarize themselves with work being done in fields that border on their particular interests.

And finally, noting the uncertainty which accompanies all philosophical debate, although aware of what philosophy has to offer, let us arise and, with Voltaire's Candide, bestir ourselves to "go and work in the garden."[2]

[1]A hint of this can be seen in the Report of the Ad Hoc Committee on Geography, The Science of Geography, pp. 62-63, whose authors advocate for geographers "a mastery of the fundamentals of two or more disciplines," where "mastery" is defined as holding a Ph.D. Although this suggestion is perhaps somewhat extreme, and although one is left to wonder as to how other disciplines would regard a Ph.D. in geography as essential to their competence, at least it repudiates the more traditional view of geography as existing in splendid isolation, and implies a closer liaison and eventual mergence of geography with other sciences.

[2]Voltaire, Candide or Optimism, trans. by John Butt (West Drayton: Penguin Books, 1947), p. 144.

Appendix

A TRANSLATION OF THE INTRODUCTION
TO KANT'S "PHYSISCHE GEOGRAPHIE"[1]

1

With respect to the whole of our knowledge, we have to direct
our attention first of all to its sources or origin, and next to
the plan of its arrangement or to its form, that is, to how this
knowledge could be ordered. Otherwise we are not in a position
to recall it with accuracy at times when we need it. Accordingly, we have to divide it, as it were, into definite disciplines
even before we obtain to knowledge itself.

2

Now, as to the sources and origin of our knowledge, it is
drawn entirely either from pure reason or from experience,
which in turn instructs reason itself.

---

[1]The translation is based on the official Rink edition of Kant's "Physische Geographie,"
GS, IX, 156-165. However, it also incorporates, by way of footnotes, suggestions
made by Erich Adickes for improving the accuracy of the text. These passages consist
either of different wordings of Kant's lectures, or of additions to the text, that occur
fairly frequently in manuscript copies examined by Adickes. Page references in these
footnotes, unless otherwise indicated, are to Adickes, Ein neu aufgefundenes Kollegheft
nach Kants Vorlesung über physische Geographie. Parts of Kant's introduction to
Physische Geographie have previously been translated by Hartshorne, The Nature of
Geography, pp. 134-135; and by Tatham, "Geography in the Nineteenth Century," pp.
38-41.

Our reason gives us knowledge of pure reason, but knowledge of experience we obtain through the senses. Since our senses do not reach beyond the world, so our knowledge of experience extends only to the present world.

However, we have two senses, an outer and an inner; thus we can view the world as the sum-total (Inbegriff) of all knowledge of experience. The world, as the object of outer sense, is nature, as the object of inner sense is soul or man.

Experience of nature and man together make up our knowledge of the world. We learn about knowledge of man in anthropology, while for knowledge of nature we are indebted to physical geography, or description of the earth. To be sure, strictly speaking, there is no sense-experience in this context, but only perceptions, which taken together would make up experience. We use that expression here really in its customary sense to mean perception.

Physical geography is thus the first part of knowledge of the world. It belongs to an idea (Idee) which is called the propaedeutic to understanding our knowledge of the world. Instruction in it still appears to be very defective. Nevertheless, it is this knowledge that is useful in all possible circumstances of life. Accordingly, it is necessary to acquaint oneself with it as a form of knowledge that may subsequently be completed and corrected by experience.

We anticipate the future experience, which we afterwards have of the world, through instruction and a general summary of this kind which gives us, as it were, a preliminary idea (Vorbegriff). We say of the person who has travelled much that he has seen the world. But more is needed for knowledge of the world than just seeing it. He who wants to profit from his journey must have a plan beforehand, and must not merely regard the world as an object of the outer senses.

The other part of knowledge of the world contains the knowledge of man. Relations with other people broaden our knowledge. Nevertheless, it is necessary to give a preliminary exercise for all future experiences of this kind, and this is what anthropology does. From it one gets acquainted with that in man which is pragmatic and not speculative. Man is not viewed physiologically but cosmologically, in order to discern the sources of phenomena.

There is a great lack of instruction as to how one may use the previously acquired knowledge, and how one may make use of it in present circumstances, or how to make our knowledge pragmatic.[2] And this is knowledge of the world.

The world is the substratum and the stage on which the play of our skills proceeds. It is the ground on which our knowledge is acquired and applied. In order that this may be exercised, and reason says it must be exercised, then the nature of the subject must be known. Without this the proper exercise of knowledge is impossible.[3]

Furthermore, we have to know the objects of our experience as a whole so that our knowledge does not form an aggregate but rather a system; in a system it is the whole that comes before the parts, whereas in an aggregate the parts are first.

It is the same with all sciences that produce in us a connection, for example the encyclopedia, where the whole appears only in the linking together of the parts. The idea (Idee) is architectonic; it creates the sciences. For example, he who wants to build a house first creates for himself an idea for the whole, from which all the parts will be derived. So our present preparation is an idea (Idee) of the knowledge of the world. Here we make for ourselves in a similar way an architectonic concept, which is a concept wherein the manifold is derived from the whole.[4]

Here, the whole is the world, the stage on which we will make all experience. Association with people and travel widen the extent of our knowledge. That association teaches us to know people, but requires much time if this end result is to be reached. But if we are prepared by instruction we already have a whole, or an essential skeleton (Inbegriff) for knowledge which teaches us how to know man. Now we are in a position to classify and arrange, within the framework, every one of the experiences we have had.[5] Through travel one broadens his

---

[2]Reading, with Adickes, "Pragmatische" for "Praktische"; notes to the "Physische Geographie," GS, IX, 515.

[3]"Until one knows man, in the way in which he has been produced and altered, the subject cannot begin" (p. 39).

[4]"I make for myself an architectonic concept, which is a concept wherein the manifold is drawn from the whole" (p. 34).

[5]"It is association and travel that widen the extent of our knowledge. Association teaches me to know man; however it requires much time to learn to know man through

knowledge of the outer world, but this is of little use unless one
has had a certain previous training through instruction. Hence,
when it is said of this or that person that he knows the world,
it is understood that he knows both man and nature.

3

Our knowledge begins with the senses. They give us the
material to which reason applies only a new[6] form. The ground
of all knowledge lies therefore in the senses and in experience.
The latter may be our own or someone else's experiences.[7]

We ought to concern ourselves only with our own experiences. However, these experiences are not sufficient to explore everything because man lives through only a small portion
of time. In addition, with respect to space, he is able to experience very little even when he travels. Although able to see
many things, he is not able to observe and perceive everything.
Therefore, we have to avail ourselves of other people's experiences. These experiences, however, have to be reliable, and
those put down in writing are preferable to those transmitted
orally.[8]

We broaden our knowledge through accounts, as if we had
lived through the whole previous world. We broaden our knowledge of the present day through news from foreign and far away
lands, as if we lived there ourselves.

But note that each foreign experience communicates itself
to us either as a story or as a description. The first is a history, the other a geography. The description of a particular
place on the earth is called topography. —The description of
a region and its characteristics is chorography. —Orography,
a description of this or that mountain range. —Hydrography,
a description of the waters.

association; however, if we are already prepared by instruction, we already have a
whole, an essential skeleton for knowledge, which teaches us to know man. Now we
are able to place every experience in its class" (p. 34).

[6]Reading, with Adickes, "neue" for "schickliche"; notes to the "Physische Geographie," GS, IX, 515.

[7]"And that may be our own or someone else's experience" (p. 34).

[8]"These, however, have to be reliable; those transmitted orally can never be as reliable as those that are in writing" (p. 35).

Remark: We are speaking here of knowledge of the world, and therefore of a description of the whole earth. The name geography is therefore taken in no other than the ordinary meaning. [9]

4

Concerning a plan for order, we have to designate a particular place to all our knowledge. We can classify our knowledge of experience either according to concepts or according to the time and place where it is actually to be found.

The classification of knowledge by concepts is the logical, that by time and space the physical classification. Through the former we obtain a system of nature (Systema Naturae), for example that of Linnaeus; through the latter a geographical description of nature.

If I say, for example, that cattle are to be classified under the species of quadrupeds, and under the kind of these animals that have cloved hooves, this is a classification that I make in my head, and therefore it is a logical classification. The system of nature is, so to speak, a register of the whole, in which I place each thing in its proper class, even though they are to be found in far-flung regions of the earth. [10]

On the other hand, following the physical classification, things are observed according to the places which they occupy on the earth. The system gives position in the classification. But the geographical description of nature shows the place in which every object on the earth is really to be found. For example, the lizard and the crocodile are fundamentally one and the same animal. The crocodile is only an immensely large lizard. But the places in which they reside are different. The crocodile lives in the Nile, the lizard on the land and even among ourselves. [11] In general, we consider here the scene of nature, the earth itself and the regions where things are really to be found. In the system of nature, however, the question is

[9]This remark was added by Rink; see his "Vorrede des Herausgebers," GS, IX, 154.

[10]"There I place each thing under its title, although they are found in various far-off and distant places in the world" (p. 35).

[11]"But they live in different places; the crocodile in the Nile, the lizard on the land" (p. 35).

not about native places but about similarities of form.

However, one should call the system of nature created up to now more correctly an aggregate of nature, because a system presupposes the idea (Idee) of a whole out of which the manifold character of things is being derived. We do not have as yet a system of nature. In the existing so-called system of this type, the objects are merely put beside each other and ordered in sequence one after the other.

We can call both history and geography, at the same time, a description, but with the difference that the former is a description of time while the latter is a description of space.[12]

History and geography enlarge our knowledge with respect to time and space. History concerns events which, under the aspect of time, have occurred one after the other. Geography concerns appearances under the aspect of space which occur simultaneously. Different names are given to the various types of objects with which geography concerns itself. Consequently, they are named physical, mathematical, political, moral, theological, literary, or commercial geography.

The history of what occurs at various times is history proper, and is nothing else than a continuous geography. Therefore, it is a great historical lack when one does not know in which location the events took place, or what the conditions were.

History therefore differs from geography only in respect to space and time. The former, as mentioned above, is an account of occurrences which have succeeded each other, and relates to time. The latter, however, is an account of occurrences which take place beside each other in space. History is a narrative, but geography is a description. Therefore we may have a description of nature, but not a history of nature.

The latter, as it is used by many, is totally incorrect. Generally, when we have only the name, we believe also that we possess the thing itself, so nobody really thinks to deliver us such a history of nature.

The history of nature contains the manifold qualities of geography, namely how things were in different epochs, but not how things are at the present time because this would be a description of nature. By contrast, if one describes the occurrences of the whole of nature as they have been through all time,

[12]"History is description according to time; description according to space is geography" (p. 35).

260

then, and only then, would one deliver a correct history of nature, as it is called. If one considers, for example, how the various breeds of dogs have come from the same root, and what changes have occurred in them in various countries and climates, and through breeding, as it has occurred in all periods, then this would be a natural history of dogs, and such a history could be delivered for every single part of nature, for example, plants, etc. But there is an awkward characteristic (Beschwerliche) here, for one would have to guess, through experiments, more than can be given in exact reports of all this. Because the history of nature is no younger than the world itself, we cannot vouch for the accuracy of our reports, not even since the invention of writing. [13] And what an immeasurable and probably far greater time lies beyond what is presented to us in recorded history!

True philosophy, however, has to follow the diversity and manifoldness of matter through all time. If the wild horses of the steppes could be tamed, they would make very enduring horses. It is noticeable that the donkey and the horse come from the same stem, and that the wild horse is the original horse because it has long ears. Also, the sheep and goat have similarities, and only the type of breeding makes a differentiation. So it is also with wine and other things.

If one were to go through the stages of nature so that one noticed what kind of changes had been gone through in all periods, then this procedure would result in a history of nature proper. [14]

The name geography therefore designates a description of nature, and at that of the whole earth. Geography and history fill up the total span of our knowledge; geography namely that of space, but history that of time.

We ordinarily assume that there is an old and a new geography, because geography has existed at all times. But which came first, history or geography? The latter is the foundation of the former, because occurrences have to refer to something.

[13]"Because the history of nature is as old as the world, and we have no reports of the time before writing began" (p. 35).

[14]"To go through the former stages of history, as they had occurred at all times, would be the history of nature. However, when one introduces this poor title, he does not endeavour to bring about a solution, since one believes what he already has" (pp. 35-36).

History is in never relenting process, but things change as well and result at times in a totally different geography. Geography therefore is the substratum. Since we have an ancient history, so naturally we must have an ancient geography.

We know best the geography of the present time. The present geography serves, among other ends, to elucidate the ancient geography by using the ancient history. However, our ordinary school geography is very deficient, although nothing is more apt for common sense than geography. Because common sense relates to experience, it is not possible for it to extend its scope to any considerable degree without a knowledge of geography. To many people, the news of the daily papers is something very insignificant. The reason for this is that they cannot locate these news items. They have no picture of the land, the ocean, and the whole surface of the earth. And yet, for instance, when the news tells something about the course of a ship through the polar sea, this is a most interesting matter because the now almost impossible hope of discovering a passage through the polar sea could bring to Europe the most important changes.[15] There is hardly another nation in which common sense and understanding is so widespread, even in the lowest classes of the population, as in the English nation. The reason for this is the newspapers, the reading of which presupposes an extended concept of the whole surface of the earth; otherwise all the news contained in the papers would leave us indifferent as we would not be able to make any application of it. The Peruvians are a simple people since they put everything that is handed to them into their mouths; they are not able to see how a more suitable use might be made of these things. Those people, who do not know how to use the news, because they don't know how to locate news items, find themselves in a situation very similar to that of these poor Peruvians.[16]

[15]"This is an interesting matter, since the invention or discovery of a passage through the polar sea might bring great changes to the whole of Europe" (p. 24).

[16]"The Peruvians are such simple people that they put everything that one gives them into their mouths, since they have no use to which to put these things; so it is also with such people who do not know how to make use of the news, for they do not know how to locate news items" (p. 36).

5

Physical geography, as a general compendium (<u>Abrisz</u>) of nature, is not only the foundation of history, but of all other possible geographies. So the main parts of each of these other geographies will have to be dealt with briefly. To these belong therefore:

1. Mathematical geography, in which the shape, size and motion of the earth, and its relationships to the solar system in which it is situated, are dealt with.

2. Moral geography, in which the diverse customs and characteristics of people of different regions are told about. For instance, when in China and especially in Japan patricide is punished as the most horrible crime, not only is the man who did the deed tortured to death in the most gruesome manner, but also his whole family is killed, and all the neighbours who lived on the same street are brought to prison. One believes that such a horrible sin cannot arise at once, but has had to develop gradually. Therefore, the neighbours ought to have foreseen this, and ought to have reported it to the authorities. In contrast, in Lappland it is a duty of love for a son to kill his father, who has been mortally wounded while hunting, with the sinew of a reindeer. Therefore, the father entrusts this sinew, at all times, to his favourite son.

3. Political geography. When the first principle of a civil society, which is a universal law as well as irresistible power, is transgressed, and if the laws extend simultaneously to the quality of the land (<u>Bodens</u>) and of its inhabitants, then political geography also belongs here inasmuch as it is wholly based on physical geography. If all the rivers of Russia flowed south, this would be of great advantage to the czardom; but they almost all flow into the polar sea. In Persia, there were, for a long time, two regents, the one residing in Ispahan, the other in Kandahar. They were incapable of overthrowing one another because they were hindered by the Kerman desert which lay between them, and which was bigger than many a sea.[17]

[17]"In Persia there were two regents, the one residing in Ispahan, the other in Kandahar. However, they were quite unable to overthrow one another. The reason (<u>Ursache</u>) for this was the great desert which lay between them, and which was larger than a sea" (p. 25).

4. Commercial geography. When a country of the earth possesses in excess the things of which another country is completely deprived, then through trade over the whole world a uniform condition will be maintained. Here, it will have to be shown why and whence one land possesses in excess that which another country goes without. More than anything else, trade has refined humanity and founded their mutual acquaintance.

5. Theological Geography. Since theological principles suffer appreciable transformation because of differences in the land (Bodens), necessary information will have to be given about this. For instance, one has only to compare the Christian religion in the Orient with the Christian religion in the West, to observe, here as there, its finer shades. This is even more sharply noticeable with religions that differ in their basic principles.

Besides this, it is necessary to notice all the deviations of nature in the differences between youth and old age, and in addition, that which is particular to each country—for instance, animals, but not the native ones, unless they are different in different countries. Thus, the nightingale does not sing as loudly in Italy as in the northern regions. On desert islands, dogs do not bark at all. In this context, one must also treat of plants, stones, herbs, mountain ranges, etc.

The usefulness of this study is very extensive. It provides a purposeful arrangement of our knowledge, serves our own entertainment, and provides rich material for social conversation.

6

Before we proceed to the treatise on physical geography proper, we must have a preconception of mathematical geography, as we have already remarked, since we will need it in our treatise only too often. Therefore, we mention here the shape, size and motion of the earth, as well as its relationship to the rest of the world-structure.

Bibliography

Ackerman, Edward A. Geography as a Fundamental Research Discipline. Department of Geography Research Paper No. 53. Chicago: Department of Geography, University of Chicago, 1958.

--------. "Where Is a Research Frontier?" Annals, Association of American Geographers, LIII, No. 4 (1963), 429-440.

Adams, Frank D. The Birth and Development of the Geological Sciences. New York: Dover, 1954. Originally published, 1938.

Ad Hoc Committee on Geography. Report of the Committee. The Science of Geography. Washington: National Academy of Sciences—National Research Council, Earth Sciences Division, 1965.

Adickes, Erich. Kants Ansichten über Geschichte und Bau der Erde. Tübingen: J. C. B. Mohr (Paul Siebeck), 1911.

--------. Untersuchungen zu Kants physischer Geographie. Tübingen: J. C. B. Mohr (Paul Siebeck), 1911.
--------. Ein neu aufgefundenes Kollegheft nach Kants Vorlesung über physische Geographie. Tübingen: J. C. B. Mohr (Paul Siebeck), 1913.

--------. Kant als Naturforscher. 2 vols. Berlin: W. de Gruyter, 1925.

Anderson, Fulton H. The Philosophy of Francis Bacon. Chicago: University of Chicago Press, 1948.

Apostel, Leo. "Towards the Formal Study of Models in the non-Formal Sciences." The Concept and the Role of the Model in Mathematics and Natural and Social Sciences. Edited by B. H. Kazemier and D. Vuysje. Dordrecht, Holland: D. Reidel, 1961.

265

Aristotle. Historia Animalium. Translated by D'Arcy W. Thompson. Vol. IV of The Works of Aristotle. Edited by W. D. Ross. Oxford: Clarendon Press, 1910.

--------. Metaphysics. Translated by W. D. Ross. The Basic Works of Aristotle. Edited by Richard McKeon. New York: Random House, 1941.

--------. Meteorologica. Translated by E. W. Webster. Vol. III of The Works of Aristotle. Edited by W. D. Ross. Oxford: Clarendon Press, 1931.

--------. Politics. Translated by Ernest Barker. Oxford: Clarendon Press, 1948.

--------. Posterior Analytics. Translated by G. R. G. Mure. The Basic Works of Aristotle. Edited by Richard McKeon. New York: Random House, 1941.

--------. (spurious). De Mundo. Translated by E. S. Forster. Vol. III of The Works of Aristotle. Edited by W. D. Ross. Oxford: Clarendon Press, 1931.

Athens Centre of Ekistics, Athens Technological Institute. International Seminar on Ekistics and the Future of Human Settlements. Pamphlet. Athens, Greece, July 20-24, 1965.

Bacon, Francis. The Advancement of Learning. London: J. M. Dent, Everyman's Library, 1915.

--------. The New Organon and Related Writings. Edited by Fulton H. Anderson. New York: Liberal Arts Press, 1960.

Bacon, Roger. The Opus Majus. Translated by Robert B. Burke. 2 vols. New York: Russell and Russell, 1962. Originally published, 1928.

Baker, J. N. L. "The Geography of Bernhard Varenius." The Institute of British Geographers, Transactions and Papers, Publication No. 21 (1955), 51-60. Reprinted, idem, The History of Geography.
--------. The History of Geography. Oxford: Basil Blackwell, 1963.
Barnard, F. M. Herder's Social and Political Thought: From Enlightenment to Nationalism. Oxford: Clarendon Press, 1965.

Barrows, Harlan H. "Geography as Human Ecology." Annals, Association of American Geographers, XIII, No. 1 (1923), 1-14.

Bates, Marston. "Human Ecology." Anthropology Today: Selections. Edited by Sol Tax. Phoenix Books. Chicago: University of Chicago Press, 1962.

Benjamin, A. Cornelius. An Introduction to the Philosophy of Science. New York: Macmillan, 1937.

Bird, Graham. Kant's Theory of Knowledge: An Outline of One Central Argument in the "Critique of Pure Reason." London: Routledge and Kegan Paul, 1962.

Black, Max. Models and Metaphors: Studies in Language and Philosophy. Ithaca: Cornell University, 1962.

Bodin, Jean. Six Books of the Commonwealth. Abridged and translated by M. J. Tooley. Oxford: Basil Blackwell, n. d.

266

Boserup, Ester. The Conditions of Agricultural Growth: The Economics of Agrarian Change under Population Pressure. London: George Allen and Unwin, 1965.

Braithwaite, Richard B. Scientific Explanation: A Study of the Function of Theory, Probability and Law in Science. New York: Harper Torchbooks, 1960.

Bridges, John H. The Life and Work of Roger Bacon: An Introduction to the "Opus Majus." London: Williams and Norgate, 1914.

Brodbeck, May. "Models, Meaning, and Theories." Symposium on Sociological Theory. Edited by Llewellyn Gross. New York: Harper and Row, 1959.

Broek, Jan O. M. Geography: Its Scope and Spirit. Columbus: Charles E. Merrill, 1965.

Bunbury, E. H. A History of Ancient Geography. 2nd edition; 2 vols. New York: Dover, 1959. Originally published, 1883.

Bunge, William. Theoretical Geography. Lund, Sweden: C. W. K. Gleerup for the Royal University of Lund, 1962.

Burgess, Ernest W., and Donald J. Bogue, eds. Contributions to Urban Sociology. Chicago: University of Chicago Press, 1964.

Burton, Ian, and Robert W. Kates. "The Perception of Natural Hazards in Resource Management." Natural Resources Journal, III, No. 3 (1964), 412-441.

Bury, J. B. The Ancient Greek Historians. New York: Dover, 1958. Originally published, 1908.

Cain, A. J. Animal Species and Their Evolution. New York: Harper Torchbooks, 1960.

Carnap, Rudolf. "Logical Foundations of the Unity of Science." International Encyclopedia of Unified Science. Edited by Otto Neurath, Rudolf Carnap, and Charles Morris. Combined edition; Vol. I, Part I. Chicago: University of Chicago Press, 1955.

--------. "Scientific Empiricism; Unity of Science Movement." The Dictionary of Philosophy. Edited by Dagobert D. Runes. London: Vision Press and Peter Owen, 1951.

Carpenter, Edmund, Frederick Varley, and Robert Flaherty. "Eskimo." Explorations, IX (1959). Toronto: University of Toronto Press, 1959.

Carpenter, Nathaniel. Geographie delineated forth in two Bookes. 2nd edition; 2 vols. Oxford: Oxford University Press, 1635.

Cassirer, Ernst. The Problem of Knowledge: Philosophy, Science, and History since Hegel. Translated by William H. Woglom and Charles W. Hendel. New Haven: Yale University Press, 1950.

--------. The Philosophy of the Enlightenment. Translated by Fritz C. A. Koelln and James P. Pettegrove. Paperback. Boston: Beacon, 1955.

--------. Rousseau, Kant and Goethe: Two Essays. Translated by James Gutmann, Paul O. Kristeller, and John H. Randall, Jr. New York: Harper Torchbooks, 1963.

Cassirer, H. W. A Commentary on Kant's "Critique of Judgment." London: Methuen, 1938.

Chorley, Richard J. "Geography and Analogue Theory." Annals, Association of American Geographers, LIV, No. 1 (1964), 127-137.

Chorley, Richard J., Antony J. Dunn, and Robert P. Beckinsale. The History of the Study of Landforms or the Development of Geomorphology. Vol. I: Geomorphology before Davis. London: Methuen, 1964.

Christaller, Walter. Central Places in Southern Germany. Translated by Carlisle W. Baskin. Englewood Cliffs: Prentice-Hall, 1966.

Clark, Andrew H. "Historical Geography." American Geography: Inventory and Prospect. Edited by Preston E. James and Clarence F. Jones. Syracuse: Syracuse University Press for the Association of American Geographers, 1954.

Clark, Colin, and Margaret Haswell. The Economics of Subsistence Agriculture. London: Macmillan, 1964.

Claval, Paul. Essai sur l'Evolution de la Géographie Humaine. Cahiers de Géographie de Besançon No. 12. Paris: Les Belles Lettres, 1964.

Cohen, Morris R. Reason and Nature: An Essay on the Meaning of Scientific Method. 2nd edition. Glencoe: Free Press, 1953.

--------. American Thought: A Critical Sketch. Glencoe: Free Press, 1954.

--------. The Meaning of Human History. 2nd edition. LaSalle: Open Court, 1961.

Collingwood, R. G. The Idea of Nature. Oxford: Clarendon Press, 1945.

--------. The Idea of History. Oxford: Clarendon Press, 1946.

Cooley, Charles Horton. "The Theory of Transportation." Sociological Theory and Social Research: Selected Papers. New York: Henry Holt, 1930. Originally published, 1894.

D'Alembert, Jean. Preliminary Discourse to the Encyclopedia of Diderot. Translated by Richard N. Schwab and Walter E. Rex. New York: Bobbs-Merrill, Library of Liberal Arts, 1963.

Dampier, William C. A History of Science and Its Relations with Philosophy and Religion. 3rd edition. New York: Macmillan, 1942.

Darby, H. C. The Theory and Practice of Geography. Liverpool: University of Liverpool Press, 1947.

--------. "The Relations of Geography and History." Geography in the Twentieth Century. Edited by Griffith Taylor. 3rd edition. London: Methuen, 1957.

Darwin, Charles. On the Origin of Species by Means of Natural Selection. 5th edition. London: John Murray, 1869.

Davis, William Morris. Geographical Essays. Edited by Douglas W. Johnson. New York: Dover, 1954. Originally published, 1909.

Dawes, Ben. A Hundred Years of Biology. London: Gerald Duckworth, 1952.

De Jong, G. Chorological Differentiation as the Fundamental Principle of Geography: An Inquiry into the Chorological Conception of Geography. Translated by H. de Jongste. Groningen, Holland: J. B. Wolters, 1962.

Dempsey, Bernard W. The Frontier Wage: The Economic Organization of Free Agents; with the Text of the Second Part of The Isolated State by Johann Heinrich von Thünen. Chicago: Loyola University, 1960.

De Vleeschauwer, Herman-J. The Development of Kantian Thought: The History of a Doctrine. Translated by A. R. C. Duncan. London: Thomas Nelson, 1962.

Dickinson, Robert E. "The Scope and Status of Urban Geography: An Assessment." Readings in Urban Geography. Edited by Harold M. Mayer and Clyde F. Kohn. Chicago: University of Chicago Press, 1959.

Dickinson, R. E., and O. J. R. Howarth. The Making of Geography. Oxford: Clarendon Press, 1933.

Doxiadis, Constantinos A. "Ekistics and Regional Science." European Congress, Zurich, 1962. Regional Science Association, Papers, X (1963), 9-46.

Dray, William. Laws and Explanation in History. Oxford: Oxford University Press, 1957.

Dryer, D. P. Kant's Solution for Verification in Metaphysics. Toronto: University of Toronto Press, 1966.

Duncan, Otis Dudley, Ray P. Cuzzort, and Beverly Duncan. Statistical Geography: Problems in Analyzing Areal Data. Glencoe: Free Press, 1961.

Durkheim, Emile. "Sociology and Its Scientific Field." Translated by Kurt H. Wolff. Emile Durkheim, 1858-1917: A Collection of Essays. Edited by Wolff. Columbus: Ohio State University, 1960.

Dury, G. H. The Face of the Earth. Harmondsworth: Penguin Books, 1959.

Evans-Pritchard, E. E. The Nuer: A Description of the Modes of Livelihood and Political Institutions of a Nilotic People. Oxford: Clarendon Press, 1940.

Ewing, A. C. A Short Commentary on Kant's "Critique of Pure Reason." Chicago: University of Chicago Press, 1938.

Fackenheim, Emil L. "Kant and Radical Evil." University of Toronto Quarterly, XXIII, No. 4 (1954), 339-353.

--------. "Kant's Concept of History." Kant-Studien, XLVIII, No. 3 (1956-57), 381-398.

Fan Shêng-Chih. On "Fan Shêng-Chih Shu": An Agriculturistic Book of China Written in the First Century B.C. Translated by Shih Shêng-Han. Peking, China: Science Press, 1959.

Farrington, Benjamin. Francis Bacon: Philosopher of Industrial Science. London: Lawrence and Wishart, 1951.

Fenneman, Nevin M. "The Circumference of Geography." Annals, Association of American Geographers, IX (1919), 3-11. Reprinted in Outside Readings in Geography. Edited by Fred E. Dohrs, Lawrence M. Sommers, and Donald R. Petterson. New York: Thomas Y. Crowell, 1958.

Firey, Walter. Man, Mind and Land: A Theory of Resource Use. Glencoe: Free Press, 1960.

Fischer, Kuno. A Critique of Kant. Translated by W. S. Hough. London: Swan Sonnenschein, Lowrey, 1888.

Geikie, Archibald. The Founders of Geology. 2nd edition. New York: Dover, 1962. Originally published, 1905.

Geiger, Rudolf. The Climate Near the Ground. Translated by Scripta Technica. 4th edition. Cambridge: Harvard University Press, 1965.

Gerland, Georg. "Immanuel Kant, seine geographischen und anthropologischen Arbeiten." Kant-Studien, X (1905), 1-43, 417-547.

Ginsberg, Morris. Sociology. Oxford: University Press, Home University Library, 1934.

Glacken, Clarence J. "Count Buffon on Cultural Changes of the Physical Environment." Annals, Association of American Geographers, L, No. 1 (1960), 1-21.

Goldsmith, Julian R. "The New Department of the Geophysical Sciences." Chicago Today, I, No. 2 (1964), 4-14.

Goudge, T. A. The Ascent of Life: A Philosophical Study of the Theory of Evolution. Toronto: University of Toronto Press, 1961.

Gregor, Mary J. Laws of Freedom: A Study of Kant's Method of Applying the Categorical Imperative in the "Metaphysik der Sitten." Oxford: Basil Blackwell, 1963.

Haeckel, Ernst. "Ueber Entwickelungsgang und Aufgabe der Zoologie." Jenäische Zeitschrift für Medicin und Naturwissenschaft, V (1870), 353-370.

Haggett, Peter. Locational Analysis in Human Geography. London: Edward Arnold, 1965.

Hakluyt, Richard. The Principal Navigations, Voyages, Traffiques and Discoveries of the English Nation. Vol. I. London: J. M. Dent, Everyman's Library, 1907.

Hartshorne, Richard. The Nature of Geography: A Critical Survey of Current Thought in the Light of the Past. Lancaster: Association of American Geographers, 1961. Originally published in Annals, Association of American Geographers, XXIX, Nos. 3 and 4 (1939).

--------. "'Exceptionalism in Geography' Re-examined." Annals, Association of American Geographers, XLV, No. 3 (1955).

--------. "The Concept of Geography as a Science of Space, from Kant and Humboldt to Hettner." Ibid., XLVIII, No. 2 (1958).

--------. Perspective on the Nature of Geography. Chicago: Rand McNally for the Association of American Geographers, 1959.

Hatt, Paul K., and Albert J. Reiss, Jr., eds. Cities and Society: The Revised Reader in Urban Sociology. Glencoe: Free Press, 1957.

Hegel, G. W. F. The Philosophy of History. Translated by J. Sibree. New York: Dover, 1956.

Heidegger, Martin. Kant and the Problem of Metaphysics. Translated by James S. Churchill. Bloomington: Indiana University, 1962.

Heidel, W. A. "Anaximander's Book, The Earliest Known Geographical Treatise." Proceedings of the American Academy of Arts and Sciences, LVI (1921), 239-288.

Hempel, Carl G. "Reasons and Covering Laws in Historical Explanation." Philosophy and History: A Symposium. Edited by Sidney Hook. New York: New York University Press, 1963.

--------. Aspects of Scientific Explanation and other Essays in the Philosophy of Science. New York: Free Press, 1965.

Hesse, Mary B. Models and Analogies in Science. Notre Dame: Notre Dame University Press, 1966.

Hesse, Richard. Ecological Animal Geography. Revised by W. C. Allee and Karl P. Schmidt. 2nd edition. New York: John Wiley, 1951.

Hettner, Alfred. "Die Entwicklung der Geographie im 19. Jahrhundert." Geographische Zeitschrift, IV (1898), 305-320.

--------. "Das Wesen und die Methoden der Geographie." Ibid., XI (1905), 545-564, 615-629, 671-686.

--------. Die Geographie: ihre Geschichte, ihr Wesen und ihre Methoden. Breslau: Ferdinand Hirt, 1927.

Hippocrates. "On Airs, Waters, and Places." Ancient Medicine and Other Treatises. Translated by Francis Adams. Chicago: Henry Regnery for the Great Books Foundation, 1949.

Hirschman, Albert O. The Strategy of Economic Development. New Haven: Yale University Press, 1958.

Hobbes, Thomas. Leviathan. Edited by Michael Oakeshott. Oxford: Basil Blackwell, n. d.

Hume, David. Essays Moral, Political and Literary. Oxford: Oxford University Press, 1963.

Huxley, Julian. The Conservation of Wild Life and Natural Habitats in Central and East Africa. Paris: UNESCO, 1961.

Ibn Khaldûn. The Muqaddimah: An Introduction to History. Translated by Franz Rosenthal. 3 vols. New York: Pantheon Books for the Bollingen Foundation, 1958.

Isard, Walter. Location and Space-Economy. Cambridge: Massachusetts Institute of Technology Press, 1956.

--------. "Regional Science, the Concept of Region, and Regional Structure." Papers and Proceedings of the Regional Science Association, II (1956), 13-26.

--------. "The Scope and Nature of Regional Science." Ibid., VI (1960), 9-34.

--------, et al. Methods of Regional Analysis: An Introduction to Regional Science. Cambridge: Massachusetts Institute of Technology Press, 1960.

James, Preston E. "Toward a Further Understanding of the Regional Concept." Annals, Association of American Geographers, XLII, No. 3 (1952), 195-222.

--------. "Geography." Encyclopaedia Britannica. 1960 edition. Vol. X, 138-152.

James, Preston E., and Clarence F. Jones, eds. American Geography: Inventory and Prospect. Syracuse: Syracuse University Press for the Association of American Geographers, 1954.

Kaminski, Willy. Über Immanuel Kants Schriften zur physischen Geographie: Ein Beitrag zur Methodik der Erdkunde. Königsberg: Hugo Jaeger, 1905.

Kant, Immanuel. Critique of Judgement. (1790.) Translated by James C. Meredith. Combined edition. Oxford: Clarendon Press, 1952.

--------. Critique of Practical Reason and Other Works on the Theory of Ethics. Translated by Thomas K. Abbott. 6th edition. London: Longmans, 1909.

--------. Critique of Pure Reason. Translated by Norman Kemp Smith. 1st edition, 1781; 2nd edition, 1787. London: Macmillan, 1950.

--------. "The Doctrine of Virtue." Part II of The Metaphysic of Morals (1797). Translated by Mary J. Gregor. New York: Harper Torchbooks, 1964.

--------. Education (1803). Translated by Annette Churton. Ann Arbor Paperbacks. Ann Arbor: University of Michigan, 1960.

--------. First Introduction to the Critique of Judgment (1789 or 1790). Translated by James Haden. New York: Bobbs-Merrill, Library of Liberal Arts, 1965.

--------. Gesammelte Schriften. 24 vols. Berlin: der Königlich Preussischen Akademie der Wissenschaften (now der Deutschen Akademie der Wissenschaften zu Berlin), 1902-1966.

--------. Introduction to Logic and Essay on the Mistaken Subtility of the Four Figures. Translated by Thomas K. Abbott. London: Longmans, Green, 1885. Reprinted, London: Vision, 1963.

--------. <u>Kant</u>. Translated and edited by Gabriele Rabel. Oxford: Clarendon Press, 1963.

--------. <u>Kant-Lexikon</u>. Edited by Rudolf Eisler. Paperback. Hildesheim, Germany: Georg Olms, 1964.

--------. <u>Kant on History</u>. Translated by Lewis W. Beck, Robert E. Anchor, and Emil L. Fackenheim. Edited by Lewis W. Beck. New York: Bobbs-Merrill, Library of Liberal Arts, 1963.

--------. <u>Kant: Philosophical Correspondence, 1759-99</u>. Translated and edited by Arnulf Zweig. Chicago: University of Chicago Press, 1967.

--------. <u>Kant's Cosmogony: As in His Essay on the Retardation of the Rotation of the Earth and His Natural History and Theory of the Heavens</u>. Translated and edited by W. Hastie. Glasgow: James Maclehose, 1900.

--------. <u>Kant's Inaugural Dissertation and Early Writings on Space</u>. Translated by John Handyside. Chicago: Open Court, 1929.

-------. <u>Lectures on Ethics</u>. Edited by Paul Menzer (1924). Translated by Louis Infield. London: Methuen, 1930.

--------. <u>Observations on the Feeling of the Beautiful and Sublime</u> (1764). Translated by John T. Goldthwait. Berkeley and Los Angeles: University of California Press, 1960.

--------. <u>On the Different Races of Man</u> (1775). Translated and edited by Earl W. Count. <u>This Is Race: An Anthology Selected from the International Literature on the Races of Man</u>. Edited by Count. New York: Henry Schuman, 1950.

--------. <u>On Philosophy in General</u> (1789 or 1790). Translated, with four introductory essays, by Humayun Kabir. Calcutta: Calcutta University Press, 1935.

--------. <u>Perpetual Peace</u> (1795). Translated by Lewis W. Beck. New York: Liberal Arts Press, 1957.

--------. <u>Prolegomena</u> and <u>Metaphysical Foundations of Natural Science</u>. Translated by Ernest Belfort Bax. London: G. Bell, Bohn's Philosophical Library, 1883.

--------. <u>Prolegomena to Any Future Metaphysics</u> (1783). Translated by Paul Carus. Revised by Lewis W. Beck. New York: Bobbs-Merrill, Library of Liberal Arts, 1950.

Kellner, L. <u>Alexander von Humboldt</u>. London: Oxford University Press, 1963.

Kettlewell, H. B. D. "Industrial Melanism." <u>Animals</u>, V, No. 20 (1965), 540-543.

Kirk, G. S., and J. E. Raven. <u>The Presocratic Philosophers: A Critical History with a Selection of Texts</u>. Cambridge: Cambridge University Press, 1957.

Klinke, Willibald. <u>Kant for Everyman</u>. Translated by Michael Bullock. London: Routledge and Kegan Paul, 1952.

Körner, S. <u>Kant</u>. Harmondsworth: Penguin Books, 1955.

Kraft, Viktor. "Die Geographie als Wissenschaft." Enzyklopädie der Erdkunde. Vol. I. Methodenlehre der Geographie. Leipzig and Vienna: Franz Deuticke, 1929.

Krüger, Gerhard. Philosophie und Moral in der Kantischen Kritik. Tübingen: J. C. B. Mohr (Paul Siebeck), 1931.

Leighly, John. "Methodologic Controversy in Nineteenth Century German Geography." Annals, Association of American Geographers, XXVIII, No. 4 (1938), 238-258.

--------, ed. Land and Life: A Selection from the Writings of Carl Ortwin Sauer. Berkeley and Los Angeles: University of California Press, 1963.

Lewis, J. R. The Ecology of Rocky Shores. London: English Universities Press, 1964.

Lewthwaite, Gordon R. "Environmentalism and Determinism: A Search for Clarification." Annals, Association of American Geographers, LVI, No. 1 (1966), 1-23.

Linné, Charles. A General System of Nature. Translated by William Turton. Vol. I. London: Lackington, Allen, 1806.

Llewellyn, Emma C., and Audrey Hawthorn. "Human Ecology." Twentieth Century Sociology. Edited by Georges Gurvitch and Wilbert E. Moore. New York: Philosophical Library, 1945.

Locke, John. Some Thoughts Concerning Education. Abridged and edited by F. W. Garforth. London: Heinemann, 1964.

Lösch, August. The Economics of Location. Translated by William H. Woglom and Wolfgang F. Stolper. 2nd edition. New Haven: Yale University, 1954.

Lovejoy, Arthur O. The Great Chain of Being: A Study of the History of an Idea. Cambridge: Harvard University Press, 1936.

Lowenthal, David. "Geography, Experience, and Imagination: Towards a Geographical Epistemology." Annals, Association of American Geographers, LI, No. 3 (1961), 241-260.

Lukermann, F. "On Explanation, Model, and Description." The Professional Geographer, XII, No. 1 (1960), 1-2.

--------. "The Role of Theory in Geographical Inquiry." Ibid., XIII, No. 2 (1961), 1-6.

--------. "The Concept of Location in Classical Geography." Annals, Association of American Geographers, LI, No. 2 (1961), 194-210.

--------. "Geography as a Formal Intellectual Discipline and the Way in Which It Contributes to Human Knowledge." The Canadian Geographer, VIII, No. 4 (1964), 167-172.

--------. Geography: de Facto or de Jure. Mimeographed. (N.p.; n. d.), 23 pp.

McKenzie, R. D. "The Scope of Human Ecology." The Urban Community. Edited by Ernest W. Burgess. Chicago: University of Chicago Press, 1926.

McNee, Robert B. "The Changing Relationships of Economics and Economic Geography." Economic Geography, XXXV, No. 3 (1959), 189-198.

McRae, Robert. The Problem of the Unity of the Sciences: Bacon to Kant. Toronto: University of Toronto Press, 1961.

Mahdi, Muhsin. Ibn Khaldûn's Philosophy of History: A Study in the Philosophic Foundation of the Science of Culture. Phoenix Books. Chicago: University of Chicago Press, 1964.

Martin, Gottfried. Kant's Metaphysics and Theory of Science. Translated by P. G. Lucas. Manchester: Manchester University Press, 1955.

Martin, S. G. "Kant as a Student of Natural Science." Immanuel Kant: Papers Read at Northwestern University on the Bicentenary of Kant's Birth. Chicago: Open Court 1925.

Maus, Heinz. A Short History of Sociology. London: Routledge and Kegan Paul, 1962.

Mayer, Harold M., and Clyde F. Kohn, eds. Readings in Urban Geography. Chicago: University of Chicago Press, 1959.

Mehlberg, Henryk. The Reach of Science. Toronto: University of Toronto Press, 1958.

Menzer, Paul. Kants Lehre von der Entwicklung in Natur und Geschichte. Berlin: Georg Reimer, 1911.

Merton, Robert K. Social Theory and Social Structure. 2nd edition. Glencoe: Free Press, 1957.

Merton, Robert K., Leonard Broom, and Leonard S. Cottrell, Jr., eds. Sociology Today: Problems and Prospects. New York: Basic Books, 1959.

Merz, John T. A History of European Thought in the Nineteenth Century. 4th edition, 4 vols. Edinburgh and London: William Blackwood, 1923-1950. Reprinted, 4 vols. New York: Dover, 1965.

Minshull, Roger. Regional Geography: Theory and Practice. Chicago: Aldine, 1967.

Mitchell, J. B. Historical Geography. London: English Universities Press, 1954.

Montefiore, A. C., and W. M. Williams. "Determinism and Possibilism." Geographical Studies, II (1955), 1-11.

Montesquieu, Baron de. The Spirit of the Laws. Translated by Thomas Nugent. New York: Hafner, Library of Classics, 1949.

Nagel, Ernest. The Structure of Science: Problems in the Logic of Scientific Explanation. New York: Harcourt, Brace and World, 1961.

Nevins, Allan. The Gateway to History. 2nd edition. Anchor Books. New York: Doubleday, 1962.

Northrop, F. S. C. "Natural Science and the Critical Philosophy of Kant." The Heritage of Kant. Edited by George T. Whitney and David F. Bowers. Princeton: Princeton University Press, 1939. Reprinted, New York: Russell and Russell, 1962.

Owen, E. E. "The Nature of Geography." Teaching Geography, No. 10 (1964). Toronto: W. J. Gage for the Education Committee of the Canadian Association of Geographers, 1964.

Owens, Joseph. A History of Ancient Western Philosophy. New York: Appleton-Century-Crofts, 1959.

Pap, Arthur. An Introduction to the Philosophy of Science. New York: Free Press of Glencoe, 1962.

Park, Robert E. "The Urban Community as a Spacial Pattern and a Moral Order." The Urban Community. Edited by Ernest W. Burgess. Chicago: University of Chicago Press, 1926.

Parsons, Talcott. The Social System. Glencoe: Free Press, 1951.

Parsons, Talcott, and Neil J. Smelser. Economy and Society: A Study in the Integration of Economic and Social Theory. Paperback. New York: Free Press, 1965.

Paton, H. J. Kant's Metaphysic of Experience: A Commentary on the First Half of the "Kritik der reinen Vernunft." 2 vols. London: George Allen and Unwin, 1936.

Paulsen, Friedrich. Immanuel Kant: His Life and Doctrine. Translated by J. E. Creighton and Albert Lefevre. 2nd edition. New York: Scribner, 1902. Reprinted, New York: Frederick Ungar, 1963.

Pearson, Lionel. Early Ionian Historians. Oxford: Clarendon Press, 1939.

Peschel, Oskar. Geschichte der Erdkunde bis auf Alexander von Humboldt und Carl Ritter. Edited by Sophus Ruge. 2nd edition. Munich: R. Oldenbourg, 1877; reprinted, Amsterdam: Meridian, 1961.

Peters, R. S., ed. Brett's History of Psychology. Abridged; 1 vol. Cambridge: Massachusetts Institute of Technology Press, 1965.

Plato. The Dialogues of Plato. Translated by B. Jowett. 2 vols. New York: Random House, 1937. Originally published, 1892.

--------. Philebus and Epinomis. Translated by A. E. Taylor. London: Thomas Nelson, 1956.

Poleman, Thomas T. The Papaloapan Project: Agricultural Development in the Mexican Tropics. Stanford: Stanford University Press, 1964.

Polunin, Nicholas. Introduction to Plant Geography and Some Related Sciences. New York: McGraw-Hill, 1960.

Polybius. The History of Polybius. Bk. 3, chaps. 57-59. Translated by A. J. Toynbee in Greek Historical Thought. New York: Mentor Books, 1952.

Popper, Karl R. The Poverty of Historicism. London: Routledge and Kegan Paul, 1957.

Ptolemy. Geographical Guide. Partly translated (Book 1, 1-5, 21-24; and Book 7, 5) by I. E. Drabkin. A Source Book in Greek Science. Edited by Morris R. Cohen and I. E. Drabkin. 2nd edition. Cambridge: Harvard University Press, 1958.

--------. Tetrabiblos. Translated by F. E. Robbins. London: William Heinemann, Loeb Classical Library, 1940.

Raven, Charles E. Natural Religion and Christian Theology. First Series. Science and Religion. Cambridge: Cambridge University Press, 1953.

Rickert, Heinrich. Science and History: A Critique of Positivist Epistemology. Translated by George Reisman. Princeton: D. van Nostrand, 1962.

Rodwin, Lloyd. "Regional Science: Quo Vadis?" Papers and Proceedings of the Regional Science Association, V (1959), 3-20.

Ross, W. D. Aristotle. 5th edition. London: Methuen, 1949.

Rousseau, Jean Jacques. The Social Contract and Discourses. Translated by G. D. H. Cole. London: J. M. Dent, Everyman's Library, 1913.

--------. Émile. Translated by Barbara Foxley. London: J. M. Dent, Everyman's Library, 1961.

Ryle, Gilbert. The Concept of Mind. London: Hutchinson, 1949.

Sarton, George. A History of Science: Ancient Science through the Golden Age of Greece. [Vol. I.] Cambridge: Harvard University Press, 1952.

--------. A History of Science: Hellenistic Science and Culture in the Last Three Centuries B.C. [Vol. II.] Cambridge: Harvard University Press, 1959.

--------. Ancient Science and Modern Civilization. New York: Harper Torchbooks, 1959.

Sartre, Jean-Paul. Being and Nothingness: An Essay on Phenomenological Ontology. Translated by Hazel E. Barnes. New York: Philosophical Library, 1956.

Schaefer, Fred K. "Exceptionalism in Geography: A Methodological Examination." Annals, Association of American Geographers, XLIII, No. 3 (1953), 226-249.

Scheidt, Walter. "The Concept of Race in Anthropology and the Divisions into Human Races from Linneus to Deniker." This Is Race: An Anthology Selected from the International Literature on the Races of Man. Edited by Earl W. Count. New York: Henry Schuman, 1950.

Schilpp, Paul A. Kant's Pre-Critical Ethics. 2nd edition. Evanston: Northwestern University Press, 1960.

Schmitthenner, H. "Alfred Hettner." Geographische Zeitschrift, XLVII (1941), 441-468. Translated by the Department of Geography and Anthropology. Baton Rouge: Louisiana State University, 1962. Mimeographed.

Schnore, Leo F. "Geography and Human Ecology." Economic Geography, XXXVII, No. 3 (1961), 207-217.

Seneca. Letter 90. "Philosophy and Progress." The Stoic Philosophy of Seneca. Translated and edited by Moses Hadas. Anchor Books. New York: Doubleday, 1958.

Sherrington, Charles. Man on His Nature. 2nd edition. Anchor Books. New York: Doubleday, 1953.

Simpson, George Gaylord. The Geography of Evolution: Collected Essays. New York: Capricorn Books, 1967.

Singer, Charles. "Biology before Aristotle." The Legacy of Greece. Edited by Richard Livingstone. Oxford: Clarendon Press, 1921.

--------. A History of Biology to about the Year 1900. 3rd edition. London and New York: Abelard-Schuman, 1959.

Sjoberg, Gideon. "Comparative Urban Sociology." Sociology Today: Problems and Prospects. Edited by Robert K. Merton, Leonard Broom, and Leonard S. Cottrell, Jr. New York: Basic Books, 1959.

Smith, Adam. An Inquiry into the Nature and Causes of the Wealth of Nations. Edited by Edwin Cannan. Modern Library. New York: Random House, 1937.

Smith, Norman Kemp. A Commentary to Kant's "Critique of Pure Reason." 2nd edition. London: Macmillan, 1923. Reprinted, New York: Humanities Press, 1962.

Sorokin, Pitirim A. Social and Cultural Mobility. Paperback. New York: Free Press of Glencoe, 1964.

Sparks, B. W. Geomorphology. London: Longmans, 1960.

Sprout, Harold, and Margaret Sprout. The Ecological Perspective on Human Affairs: With Special Reference to International Politics. Princeton: Princeton University Press, 1965.

Stahl, William H. Roman Science: Origins, Development, and Influence to the Later Middle Ages. Madison: University of Wisconsin, 1962.

Strabo. The Geography of Strabo. Translated by H. L. Jones. 8 vols. London: William Heinemann, Loeb Classical Library, 1917-1932.

--------. Des Strabo allgemeine Erdbeschreibung. Translated by Abraham J. Penzel. 4 vols. Lemgo: Menerschen Buchhandlung, 1775-1777.

Strahler, Arthur N. Physical Geography. 2nd edition. New York: John Wiley, 1960.

Strasser, Stephan. Phenomenology and the Human Sciences: A Contribution to a New Scientific Ideal. Pittsburgh: Duquesne University, 1963.

Stuckenberg, J. W. H. The Life of Immanuel Kant. London, 1882.

Tatham, George. "Geography in the Nineteenth Century." Geography in the Twentieth Century. Edited by Griffith Taylor. 3rd edition. London: Methuen, 1957.

--------. "Environmentalism and Possibilism." Ibid.

Taylor, E. G. R. Tudor Geography, 1485-1583. London: Methuen, 1930.

--------. Late Tudor and Early Stuart Geography, 1583-1650. London: Methuen, 1934.

Taylor, Griffith. "Geography the Correlative Science." Canadian Journal of Economics and Political Science, I, No. 4 (1935), 535-550.

--------. Our Evolving Civilization: An Introduction to Geopacifics. Toronto: University of Toronto Press, 1946.

--------. "Introduction: The Scope of the Volume." Geography in the Twentieth Century. Edited by G. Taylor. 3rd edition. London: Methuen, 1957.

Thomas, William L., Jr., ed. Man's Role in Changing the Face of the Earth. Chicago: University of Chicago Press for the Wenner-Gren Foundation, 1956.

Thomson, J. Oliver. History of Ancient Geography. Cambridge: Cambridge University Press, 1948. Reprinted, New York: Biblo and Tannen, 1965.

Thornbury, William D. Principles of Geomorphology. New York: John Wiley, 1954.

Toulmin, Stephen. The Philosophy of Science: An Introduction. New York: Harper Torchbooks, 1960.

Tozer, H. F. A History of Ancient Geography. Revised by M. Cary. Cambridge: Cambridge University Press, 1935. Reprinted, New York: Biblo and Tannen, 1964.

Ullman, Edward. "A Theory of Location for Cities." Readings in Urban Geography. Edited by Harold M. Mayer and Clyde F. Kohn. Chicago: University of Chicago Press, 1959.

Van Paassen, C. The Classical Tradition of Geography. Translated by C. M. Reith-Aerssens, B. J. Wevers, and R. R. Symonds. Groningen, Holland: J. B. Wolters, 1957.

--------. "Carl Ritter Anno 1959." Tijdschrift van het Koninklijk Nederlandsch Aardrijkskundig Genootschap, LXXVI, No. 4 (1959), 327-351.

Varenius, Bernhard. Géographie Générale. Revue par Isaac Newton. Augmentée par Jacques Jurin. Paris, 1755.

Vico, Giambattista. The New Science of Giambattista Vico. Translated by Thomas G. Bergin and Max H. Fisch. 3rd edition. Anchor Books. New York: Doubleday, 1961.

Vidal de la Blache, Paul. "Des caractères distinctifs de la géographie." Annales de Géographie, XXII (1913), 289-299.

--------. Principles of Human Geography. Translated by Millicent T. Bingham. Edited by Emmanuel de Martonne. London: Constable, 1926.

Von Humboldt, Alexander. Cosmos. [Kosmos.] Translated by E. C. Otté. 5 vols. London: H. G. Bohn, Bohn's Scientific Library, 1849-1863.

Von Humboldt, Wilhelm. Humanist without Portfolio: An Anthology of the Writings of Wilhelm von Humboldt. Translated and edited by Marianne Cowan. Detroit: Wayne State University, 1963.

Von Thünen, Johann Heinrich. The Isolated State. Translated by Carla M. Wartenberg. Edited by Peter Hall. Oxford: Permagon Press, 1966. See also under Dempsey.

Wagner, Philip L. The Human Use of the Earth. Glencoe: Free Press, 1960.

Watkins, J. W. N. "Ideal Types and Historical Explanation." Readings in the Philosophy of Science. Edited by Herbert Feigl and May Brodbeck. New York: Appleton-Century-Crofts, 1953.

Weber, Alfred. Theory of the Location of Industries. Translated by Carl J. Friedrich. Chicago: University of Chicago Press, 1929.

Weber, Max. The Theory of Social and Economic Organization. Translated by A. R. Henderson and Talcott Parsons. London: William Hodge, 1947.

--------. The Methodology of the Social Sciences. Translated by Edward A. Shils and Henry A. Finch. Glencoe: Free Press, 1949.

Weldon, T. D. Kant's "Critique of Pure Reason." 2nd edition. Oxford: Clarendon Press, 1958.

Whittlesey, Derwent. The Earth and the State. 2nd edition. New York: Henry Holt, 1944.

--------. "The Regional Concept and the Regional Method." American Geography: Inventory and Prospect. Edited by Preston E. James and Clarence F. Jones. Syracuse: Syracuse University Press for the Association of American Geographers, 1954.

Windelband, Wilhelm. Theories in Logic. Translated by Thomas P. Kiernan. New York: Citadel, 1961.

Wolf, A. A History of Science, Technology and Philosophy in the 16th and 17th Centuries. 2nd edition; 2 vols. New York: Harper Torchbooks, 1959.

--------. A History of Science, Technology, and Philosophy in the 18th Century. Revised by D. McKie. 2nd edition; 2 vols. New York: Harper Torchbooks, 1961.

Wolff, Robert P. Kant's Theory of Mental Activity: A Commentary on the Transcendental Analytic of the "Critique of Pure Reason." Cambridge: Harvard University Press, 1963.

Wooldridge, S. W., and W. Gordon East. The Spirit and Purpose of Geography. 2nd edition. London: Hutchinson, University Library, 1958.

Wright, J. K. "Some British 'Grandfathers' of American Geography." Geographical Essays in Memory of Alan G. Ogilvie. Edited by R. Miller and J. Wreford Watson. London: Thomas Nelson, 1959.

280

Xenophon. Ways and Means: A Pamphlet on Revenues. Translated by Henry G. Dakyns. The Greek Historians. Vol. II. Edited by Francis R. B. Godolphin. New York: Random House, 1942.

Zeller, Eduard. Outlines of the History of Greek Philosophy. Translated by L. R. Palmer. Revised by Wilhelm Nestle. 13th edition. London: Routledge and Kegan Paul, 1931.

Zimmermann, Erich W. World Resources and Industries: A Functional Appraisal of the Availability of Agricultural and Industrial Materials. 2nd edition. New York: Harper, 1951.